The Slide Rule, Electronic Hand Calculator, and Metrification in Problem Solving

Third Edition

George C. Beakley
Arizona State University

H. W. Leach
Bell Helicopter Company

Macmillan Publishing Co., Inc.
New York
Collier Macmillan Publishing
London

Earlier edition entitled *The Slide Rule* copyright 1953 by H. W. Leach and George C. Beakley. Earlier edition entitled *The Slide Rule And Technical Problem Solving* © 1963 by Macmillan Publishing Co., Inc. Earlier edition entitled *The Slide Rule and Its Use in Problem Solving* copyright © 1969 by George C. Beakley and H. W. Leach.

Selected material has been reprinted from *Engineering: An Introduction to a Creative Profession,* Second Edition, copyright © 1972 by George C. Beakley and H. W. Leach, and from *Introduction to Engineering Design and Graphics,* copyright © 1973 by George C. Beakley.

Macmillan Publishing Co., Inc.
866 Third Avenue, New York, New York 10022

Collier-Macmillan Canada, Ltd.

Library of Congress Cataloging in Publication Data

Beakley, George C
The slide rule, electronic hand calculator, and metrification in problem solving.

Second ed. published in 1969 under title: The slide rule and its use in problem solving.
Includes index.
1. Slide-rule. 2. Calculating-machines. 3. Metric system. 4. Problem solving.
I. Leach, H. W., joint author. II. Title.
QA73.L36 1975 510'.28 74-22262
ISBN 0-02-307220-2

Printing: 1 2 3 4 5 6 7 8 Year: 5 6 7 8 9 0

preface

Today is a time of change. This generation will have the unique experience of participating in the alteration of a system of measurements that has been used in our country for hundreds of years. Of lesser import, but of great significance to students of mathematics, science, engineering, and technology, are the revolutionary advances that have been made in the design and manufacture of computers and small electronic computational machines which greatly enhance the speed and accuracy of analytical procedures. For over 200 years the slide rule has been an international "workhorse instrument" for individual calculation purposes, and because of its low cost it will continue in dominance for a number of years in the future. However, the widespread adoption of small hand-held solid-state calculators is only a matter of time.

Previous editions of this text have emphasized primarily the slide rule and technical problem solving. These areas have been amplified and strengthened in this third edition. In addition, extensive chapters have been added on the use of electronic hand calculators and metrification. The new international SI system is emphasized, but numerous problem solving exercises are included for both metric and English units.

The subject matter in this text is arranged so that it is suitable both for classroom use and for individual study. In presenting the text material, the authors have stressed the practical application of mathematical tools to help the student grasp the role of mathematical skills in problem solving situations.

The sections on the slide rule and electronic hand calculators have been arranged so that the student can follow the examples shown for each operation, and then check learning progress with a set of problem situations for which answers are given. The sections on the problem solving process and general problem solving are designed to present situations involving the application of basic physical principles. The student is encouraged not only to use step-by-step reasoning in analyzing each problem but also to present the solution in a neat and orderly manner.

We wish to express appreciation to and acknowledge the assistance of several manufacturers of electronic hand calculators. These companies, Hewlett-Packard, Texas Instruments Incorporated, Unicom Systems, Bowmar Consumer Products Division, and Sharp Electronics Corporation, made available to the authors photographs and instructional materials for use in Chapter Three.

We are especially interested in learning the opinions of those who read this book concerning its utility and serviceability in meeting the needs for which it was written. Improvements that are suggested will be considered for incorporation in later editions.

G. C. B.
H. W. L.

contents

1

presentation of engineering and scientific calculations

FORMAT

In problem solving, both in school and in industry, considerable importance is attached to a proper analysis of the problem, to a logical recording of the problem solution, and to the overall professional appearance of the finished calculations. Neatness and clarity of presentation are distinguishing marks of the engineer's work. Students should strive always to practice professional habits of problem analysis and to make a conscious effort to improve the appearance of each paper, whether it is submitted for grading or is included in a notebook.

The computation paper used for most calculations is 8½ by 11 inches in size, with lines ruled both vertically and horizontally on the sheet. Usually these lines divide the paper into five squares per inch, and the paper is commonly known as cross-section paper or engineering calculation paper. Many schools use paper that has the lines ruled on the reverse side of the paper so that erasures will not remove them. A fundamental principle to be followed is that the problem work shown on the paper should not be crowded and that all steps of the solution should be included.

Engineers use slant or vertical lettering (see Figure 1–1); either is acceptable as long as there is no mixing of the two forms. The student should not be discouraged if he finds that he cannot letter with great speed and dexterity at first. Skills in making good letters improve with hours of patient practice. Use a well-sharpened H or 2H pencil and follow the sequence of strokes recommended in Figure 1–1.

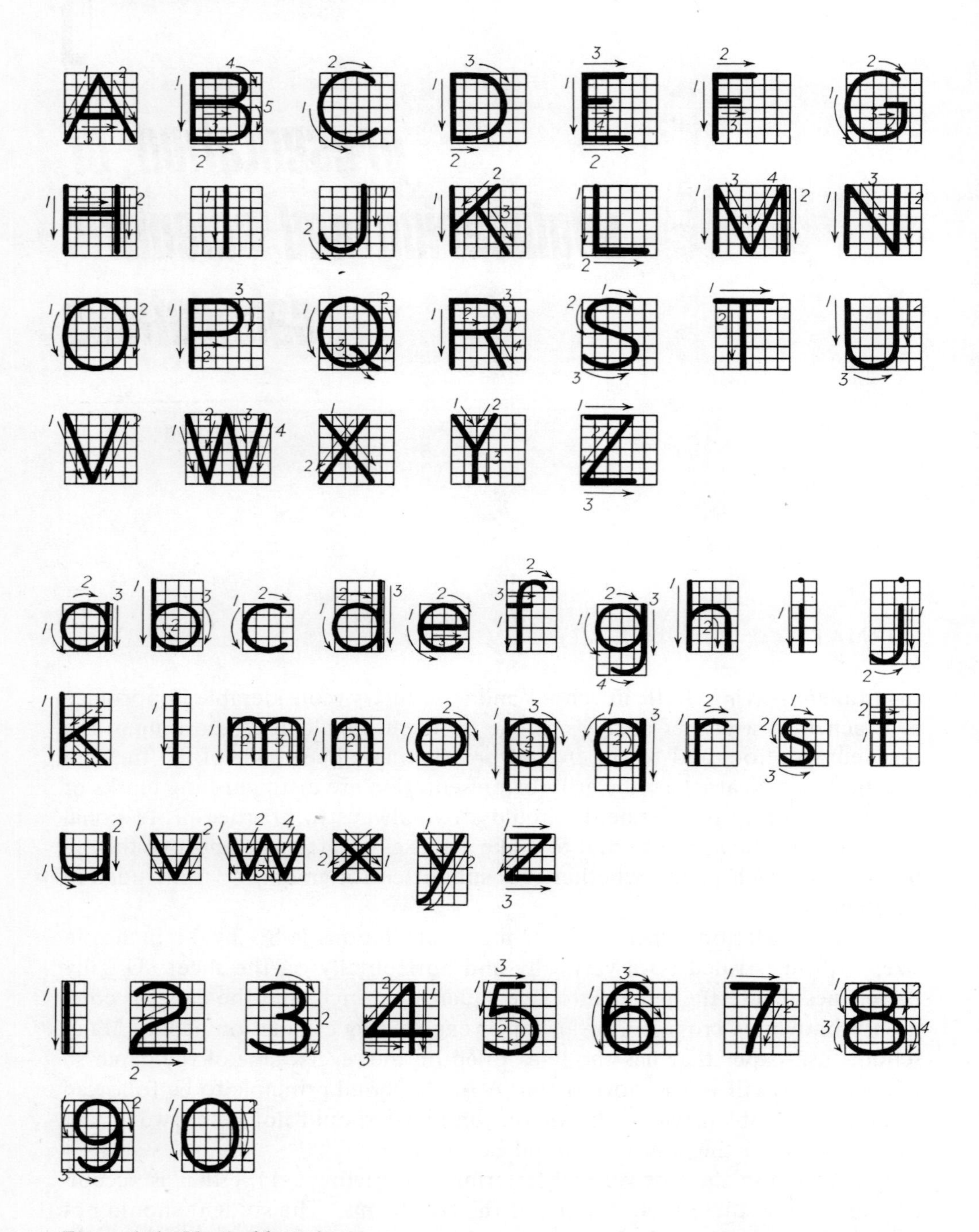

Figure 1–1. Vertical lettering.

Several styles of model problem sheets are shown in Figures 1–2 to 1–5. Notice in each sample that an orderly sequence is followed in which the known data are given first. The data are followed by a brief statement of the requirements, and then the engineer's solution.

Problem 1 (Algebra) Smith, Bill

a. $(x^n)^4 (x^2) = \underline{\underline{x^{4n+2}}}$

b. $\frac{x^7}{x^2} = \underline{\underline{x^5}}$

c. $(y^4)(y^3) = \underline{\underline{y^7}}$

Problem 8 (Logarithms)

GIVEN:

a. (35) (6) = Ans.

b. $\frac{(400)}{(75)}$ = Ans.

SOLUTION:

a. log ans. = log 35 + log 6
log 35 = 1.5441
log 6 = 0.7782
log ans. = 2.3223
ans. = 210

b. log ans. = log 400 − log 75
log 400 = 2.6021
log 75 = 1.8751
log ans. = 0.7270
ans. = 5.33

Figure 1–2. Model problem sheet, style A. This style shows a method of presenting short, simple exercises.

When the problem solution is finished, the paper may be folded and endorsed on the outside or may be submitted flat in a folder. Items that appear on the endorsement should include the student's name, and the course, section, date, problem numbers, and any other prescribed information. An example of a paper that has been folded and endorsed is shown in Figure 1–6.

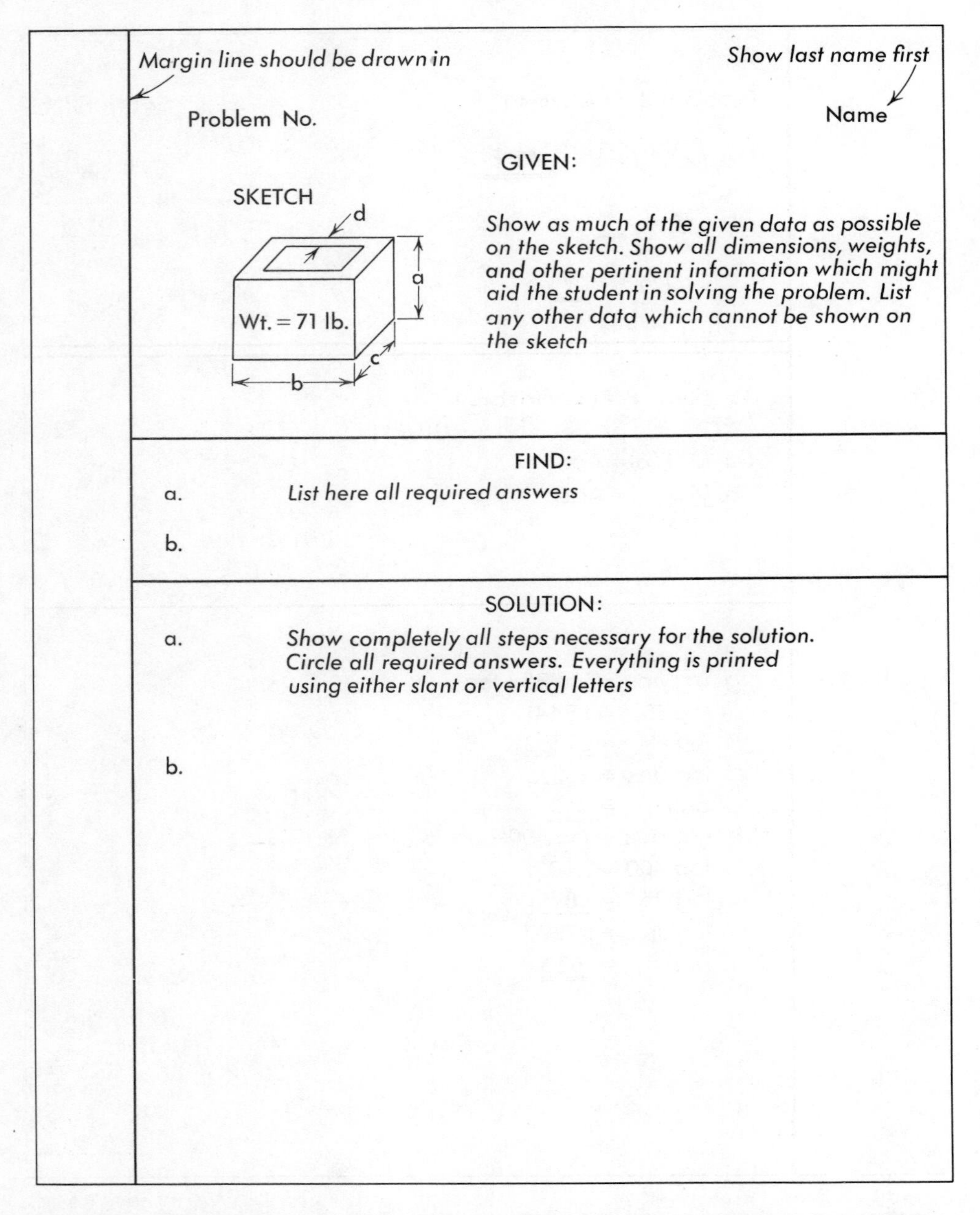

Figure 1–3. Model problem sheet, style B. This style shows a general form which is useful in presenting the solution of mensuration problems.

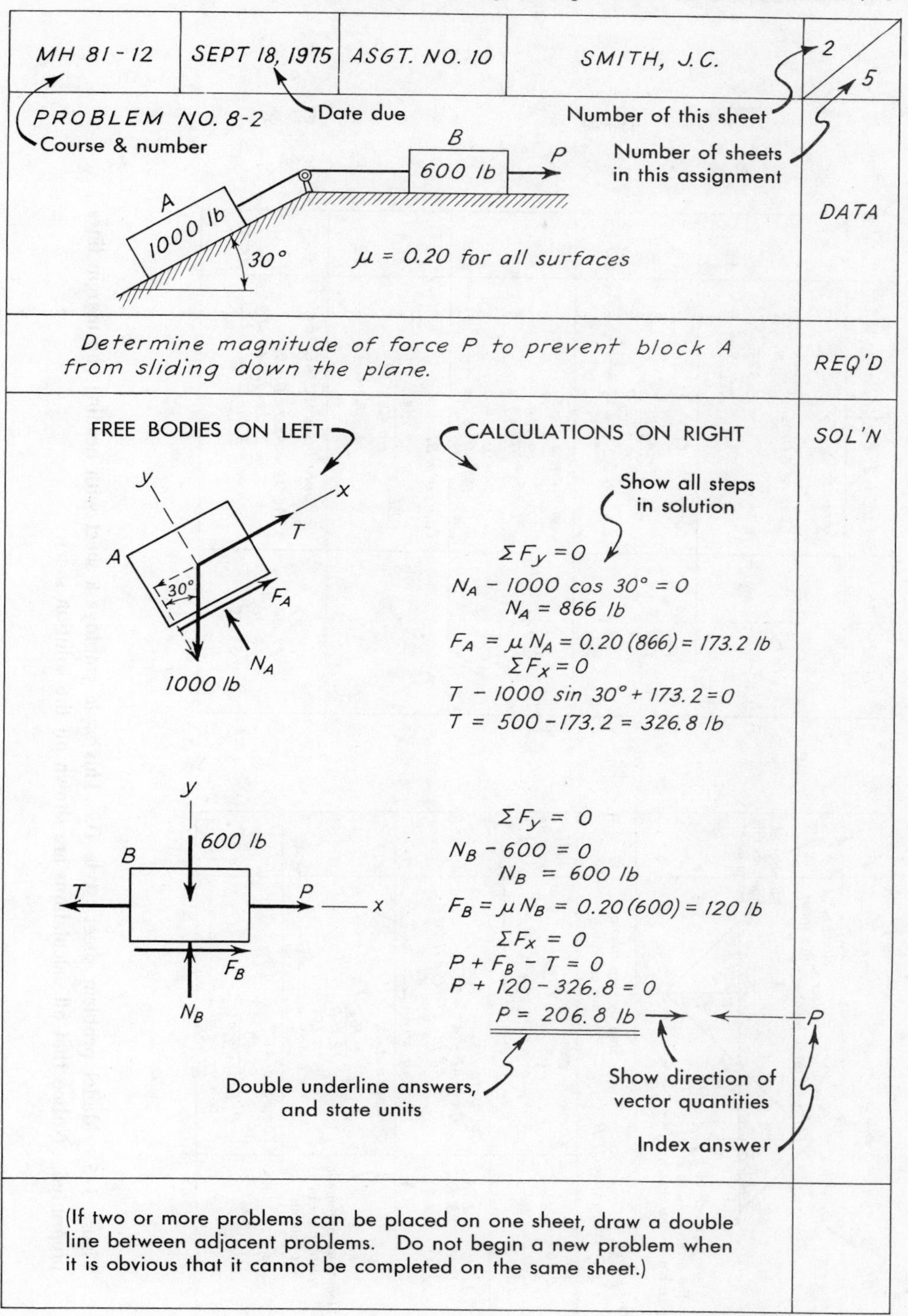

Figure 1–4. Model problem sheet, style C. This style shows a method of presenting stated problems. Notice that all calculations are shown on the sheet and that no scratch calculations on other sheets are used.

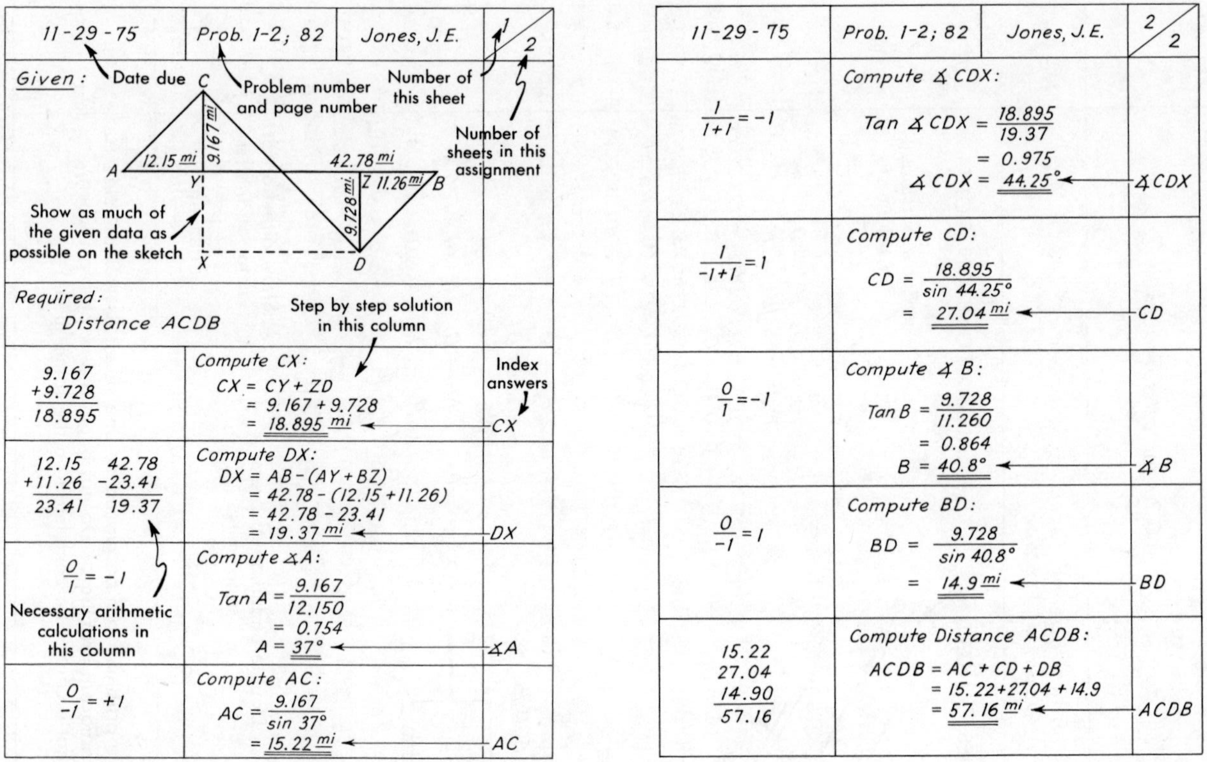

11-29-75	Prob. 1-2; 82	Jones, J.E.	1/2
Given: Date due Show as much of the given data as possible on the sketch	C, A, Y, X, D, Z, B 12.15 mi, 9.167 mi, 42.78 mi, 11.26 mi, 9.728 mi Problem number and page number	Number of this sheet Number of sheets in this assignment	
Required: Distance ACDB	Step by step solution in this column		
9.167 +9.728 18.895	Compute CX: $CX = CY + ZD$ $= 9.167 + 9.728$ $= \underline{\underline{18.895\ mi}}$		Index answers CX
12.15 42.78 +11.26 −23.41 23.41 19.37	Compute DX: $DX = AB - (AY + BZ)$ $= 42.78 - (12.15 + 11.26)$ $= 42.78 - 23.41$ $= \underline{\underline{19.37\ mi}}$		DX
$\frac{0}{1} = -1$ Necessary arithmetic calculations in this column	Compute ∡A: $\tan A = \frac{9.167}{12.150}$ $= 0.754$ $A = \underline{\underline{37°}}$		∡A
$\frac{0}{-1} = +1$	Compute AC: $AC = \frac{9.167}{\sin 37°}$ $= \underline{\underline{15.22\ mi}}$		AC

11-29-75	Prob. 1-2; 82	Jones, J.E.	2/2
$\frac{1}{1+1} = -1$	Compute ∡CDX: $\tan ∡CDX = \frac{18.895}{19.37}$ $= 0.975$ $∡CDX = \underline{\underline{44.25°}}$		∡CDX
$\frac{1}{-1+1} = 1$	Compute CD: $CD = \frac{18.895}{\sin 44.25°}$ $= \underline{\underline{27.04\ mi}}$		CD
$\frac{0}{1} = -1$	Compute ∡B: $\tan B = \frac{9.728}{11.260}$ $= 0.864$ $B = \underline{\underline{40.8°}}$		∡B
$\frac{0}{-1} = 1$	Compute BD: $BD = \frac{9.728}{\sin 40.8°}$ $= \underline{\underline{14.9\ mi}}$		BD
15.22 27.04 14.90 57.16	Compute Distance ACDB: $ACDB = AC + CD + DB$ $= 15.22 + 27.04 + 14.9$ $= \underline{\underline{57.16\ mi}}$		ACDB

Figure 1-5. Model problem sheet, style D. This style employs a sheet with heading and margin lines preprinted. Notice that all calculations are shown on the solution sheet.

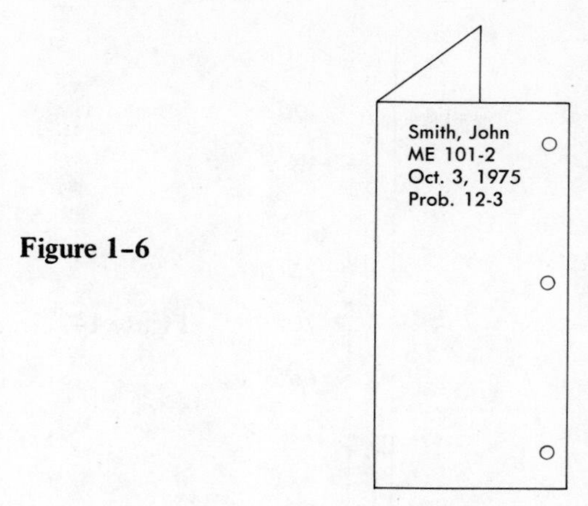

Figure 1–6

Scientific presentation of measured data

Since measured data inherently are not exact, it is necessary that methods of manipulating data be examined so that information derived therefrom can be evaluated properly. It should be obvious that the diameter of a saucepan and the diameter of a diesel engine piston, although each may measure about 6 in., usually will be measured with different accuracies. Also a measurement of the area of a large ranch which is valued at $50 per acre would not be made as accurately as a measurement of a piece of commercial property that is valued at $1000 per square foot. In order to describe the accuracy of a single measurement, it can be given in terms of a set of significant figures.

Significant figures

A significant figure in a number can be defined as a figure that may be considered reliable as a result of measurements or of mathematical computations. In making measurements, it is customary to read and record all figures from the graduations on the measuring device and to include one estimated figure which is a fractional part of the smallest graduation. Any instrument can be assumed to be accurate *only* to one half of the smallest scale division that has been marked by the manufacturer. All figures read are considered to be significant figures. For example, if we examine the sketch of the thermometer in Figure 1–7, we see that the mercury column, represented in the sketch by a vertical line, lies between 71° and 72°. Since the smallest graduation is 1°, we should record 71° and include an estimated 0.5°. The reading would then be recorded as 71.5° and would contain three significant figures.

As another example, suppose that it is necessary to record the voltmeter reading shown in Figure 1–8. The needle obviously rests between the graduations of 20 and 30 volts. A closer inspection shows that its location can be more closely determined as being between 25 and 26 volts. However, this is

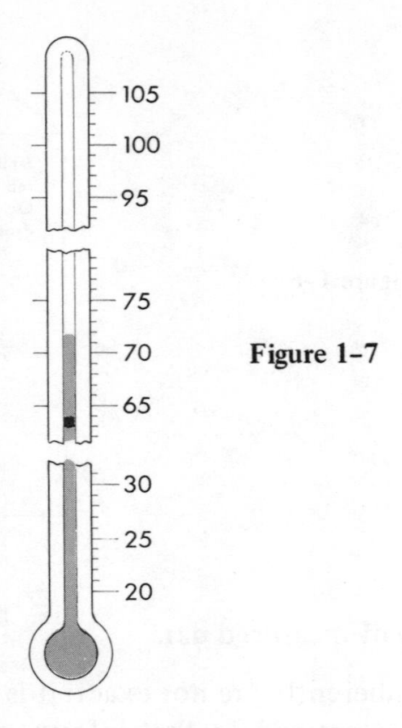

Figure 1–7

the extent of the aid which we can get from the individual graduations. Any further refinement must be accomplished by eye.[1] Since the scale of the voltmeter is calibrated to the nearest volt, we can estimate the reading to the nearest half volt—in this case 25.5 volts. An attempt to obtain a more precise reading (such as 25.6 or 25.7) would result only in false accuracy, as discussed below.

The designated digits, together with one doubtful digit, are said to be "significant figures." In reading values previously recorded, assume that only one doubtful digit has been recorded. This usually will be the last digit retained in any recorded measurement.

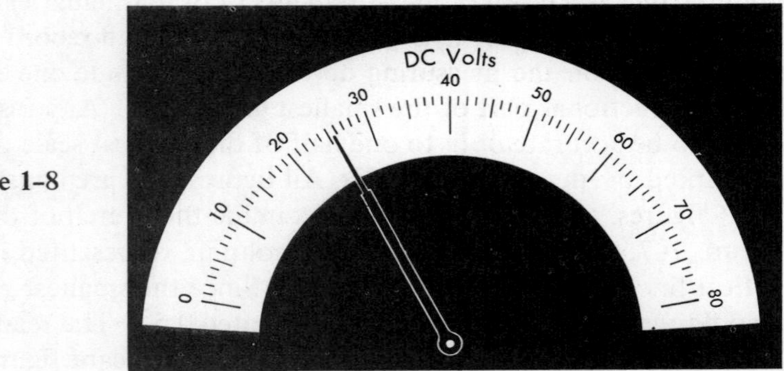

Figure 1–8

[1]In most cases, estimation by eye (beyond the precision obtainable from the graduations) is acceptable. It should be recognized that this final subdivision (by eye) will give doubtful results.

False accuracy

In analysis of engineering problems one must prevent false accuracy from appearing in the calculations. False accuracy occurs when data are manipulated without regard to their degree of precision. For example, it may be desirable to find the sum of three lengths, each having been measured with a different type of instrument. These lengths might have been recorded in tabular form (rows and columns) as:

		Columns a b c d e f g	
First Measurement:	Row A	1 5 7 . 3 9	±0.02 ft.
Second Measurement:	Row B	1 8 . 0 2 5	±0.001 ft.
Third Measurement:	Row C	8 5 3 .	±2 ft.
		1 0 2 8 . 4 1 5	(By regular addition)

Although the sum of the columns would be 1028.415, it would not be proper to use this value in other calculations. Since the last measurement (Row C) could vary from 851 to 855 (maximum variation in Column d), it would be trivial to include the decimal numbers in Rows A and B in the sum. The final answer should be expressed as 1028 ± 2, or merely 1028. In this case the last digit (8) is of doubtful accuracy.

In the tabulation of data (readings from meters, dials, gages, verniers, scales, etc.), only one doubtful digit may be retained for any measurement. In the preceding example, the doubtful digits are 9 (Row A), 5 (Row B), and 3 (Row C). The example also shows that when numbers are added, the sum should not be written to more digits than the digit under the first column which has a doubtful number.

Scientific notation

The decimal point has nothing to do with how many significant figures there are in a number, and therefore it is impossible to tell the number of significant figures if written as 176,000., 96000., or 1000. This doubt can be removed by the following procedure:

1. Move the decimal point to the left or right until a number between 1 and 10 remains. The number resulting from this process should contain *only* significant figures.

2. This remaining number must now be multiplied by a power of ten, $(10)^{\text{number of decimal moves}}$. If the decimal is moved to the left, the power of 10 is positive.

Example Express the number 1756000 to five significant figures:

1ₓ7 5 6 0 0 0. (Move the decimal point to the left to get a number between 1 and 10.)
  6 5 4 3 2 1

Answer **$(1.7560)(10)^6$** (The power of 10 is the number of decimal moves.)

Note Only the five significant figures remain to be multiplied by the power of 10.

Example Express the number 0.016900 to three significant figures:

0 . 0 1 x 6 9 (Move the decimal point to the right to get a number between 1 and 10.)
1 2

Answer **$(1.69)(10)^{-2}$** (The power of 10 is the number of decimal moves and is negative in sign.)

Note The three significant figures remain to be multiplied by the power of 10.

Examples of significant figures:

385.1	four significant figures
38.51	four significant figures
0.03851	four significant figures
3.851×10^7	four significant figures
7.04×10^{-4}	three significant figures
25.5	three significant figures
0.051	two significant figures
0.00005	one significant figure
27,855	five significant figures
8.91×10^4	three significant figures
2200	May have two, three, or four significant figures depending on the accuracy of the measurement that obtained the number. Where such doubt may exist, it is better to write the number as 2.2×10^3 to show two significant figures; or as 2.20×10^3 to show three significant figures.
55	two significant figures
55.0	three significant figures. The zero is significant in this case, since it is not otherwise needed to show proper location of the decimal point.

In engineering computations it is necessary to use standard computed constants, such as π (3.14159265 . . .) and ϵ (2.71828 . . .). It is feasible to simplify these values to fewer significant figures, since most calculations will be done on the slide rule where five, six, and seven significant figures are impossible to read. Usually three or four significant figures are sufficient, but this may vary somewhat with the nature of the problem. Since we do not need a large number of significant figures, let us examine some rules concerning "rounding off" the excess figures which need not be used in a given calculation.

Retention of significant figures

1. In recording measured data, only one doubtful digit is retained, and it is considered to be a significant figure.
2. In dropping figures which are not significant, the last figure retained should be increased by 1 if the first figure dropped is 5 or greater.
3. In addition and subtraction, do not carry the result beyond the first column which contains a doubtful figure.
4. In multiplication and division, carry the result to the same number of significant figures that there are in the quantity entering into the calculation which has the least number of significant figures.

1-1. Determine the proper value of X for each problem.

a. $0.785 = 7.85(10^x)$
b. $0.005066 = 5.066(10^x)$
c. $6.45 = 64.5(10^x)$
d. $10.764 = 10764(10^x)$
e. $1973 = 0.01973(10^x)$
f. $0.3937 = 3937000(10^x)$
g. $30.48 = 0.03048(10^x)$
h. $2.54 = 254(10^x)$
i. $1000 = 10(10^x)$
j. $0.001 = 1(10^x)$
k. $44.2 = 0.442(10^x)$
l. $0.737 = 73.7(10^x)$
m. $1.093 = 10930(10^x)$
n. $4961 = 0.4961(10^x)$

2

the slide rule

The slide rule is not a modern invention although its extensive use in business and industry has been common only in recent years. Since the slide rule is a mechanical device whereby the logarithms of numbers may be manipulated, the slide rule of today was made possible over three and a half centuries ago with the invention of logarithms by John Napier, Baron of Merchiston in Scotland. Although Napier did not publicly announce his system of logarithms until 1614, he had privately communicated a summary of his results to Tycho Brahe, a Danish astronomer in 1594. Napier set forth his purpose with these words:

> Seeing there is nothing (right well beloved Students of Mathematics) that is so troublesome to mathematical practice, nor doth more molest and hinder calculators, than the multiplications, divisions, square and cubical extractions of great numbers, which besides the tedious expense of time are for the most part subject to many slippery errors, I began therefore to consider in my mind by what certain and ready art I might remove those hindrances.

In 1620 Edmund Gunter, Professor of Astronomy at Gresham College, in London, conceived the idea of using logarithm scales that were constructed with antilogarithm markings for use in simple mathematical operations. William Oughtred, who lived near London, first used "Gunter's logarithm scales" in 1630 in sliding combination, thereby creating the first slide rule. Later he also placed the logarithm scales in circular form for use as a "circular form for use as a "circular slide rule."

Sir Isaac Newton, John Warner, John Robertson, Peter Roget, and Lieutenant Amédée Mannheim further developed these logarithmic scales until there exist today many types and shapes of rules. Basically all rules of modern manufacture are variations of a general type of construction that utilizes sliding scales and a movable indicator. The principles of operation are the same and they are not difficult to master.

DESCRIPTION OF THE SLIDE RULE

The slide rule consists of three main parts, the "body," the "slide," and the "indicator" (see Figure 2–1). The "body" of the rule is fixed; the "slide" is the middle sliding portion; and the "indicator," which may slide right or left on the body of the rule, is the transparent runner. A finely etched line on each side of the indicator is used to improve the accuracy in making settings and for locating the answer. This line is referred to as the "hairline."

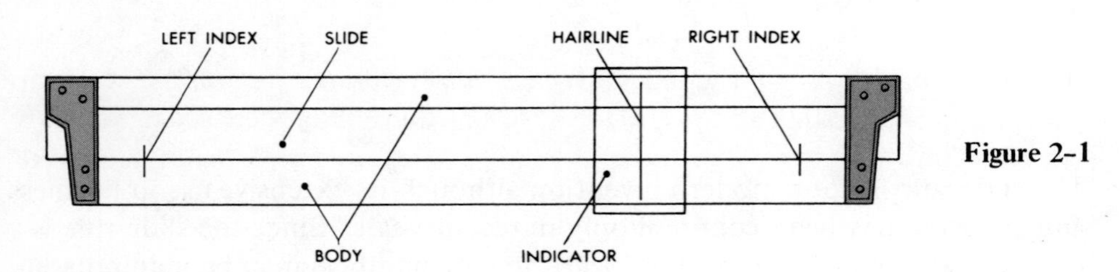

Figure 2–1

The mark opposite the primary number 1 on the C and D scale is referred to as the "index" of the scale. An examination of the C and D scales indicates that each scale has two indexes: one at the left end (called the left index") and one at the right end (called the "right index").

Regardless of the manufacturer or the specific model of slide rule that may be used, the principles of operation are the same. The nomenclature used here is general although some specific references are made to the Deci-Lon (Keuffel & Esser Co.), the Model 10,000 KOH–I–NOOR, Inc., the Versalog (Frederick Post Co.), the Maniphase Multiplex (Eugene Dietzgen Co.), and the Model N4 (Pickett, Inc.) rules. These models are those most frequently used by engineers, scientists, and technicians.

CARE OF THE SLIDE RULE

The slide rule is a precision instrument and should be afforded reasonable care in order to preserve its accuracy. Modern rules stand up well under normal usage, but dropping the rule or striking objects with it will probably impair its accuracy.

In use, the rule may collect dirt under the glass of the indicator. Inserting a piece of paper under the glass and sliding the indicator across it will frequently dislodge the dirt without necessitating the removal of the indicator glass from the frame. If the glass has to be removed for cleaning, it should be realigned when replaced, using the techniques described below.

The rule should never be washed with abrasive materials, alcohol, or other solvents, since these may remove markings. If the rule needs to be cleaned, it may be wiped carefully with a damp cloth, but the excessive use of water should be avoided because it will cause wooden rules to warp.

The metal-frame rules are not subject to warping due to moisture changes, but they must be protected against blows which would bend them or otherwise throw them out of alignment. A light layer of lubricant of the type specified by the manufacturer of the metal rule will increase the ease with which the working parts move. This is particularly important during the "breaking in" period of the new rule.

MANIPULATION OF THE RULE

Some techniques in manipulation of the rule have been found to speed up the setting of the slide and indicator. Two of these suggested procedures are described in the following paragraphs.

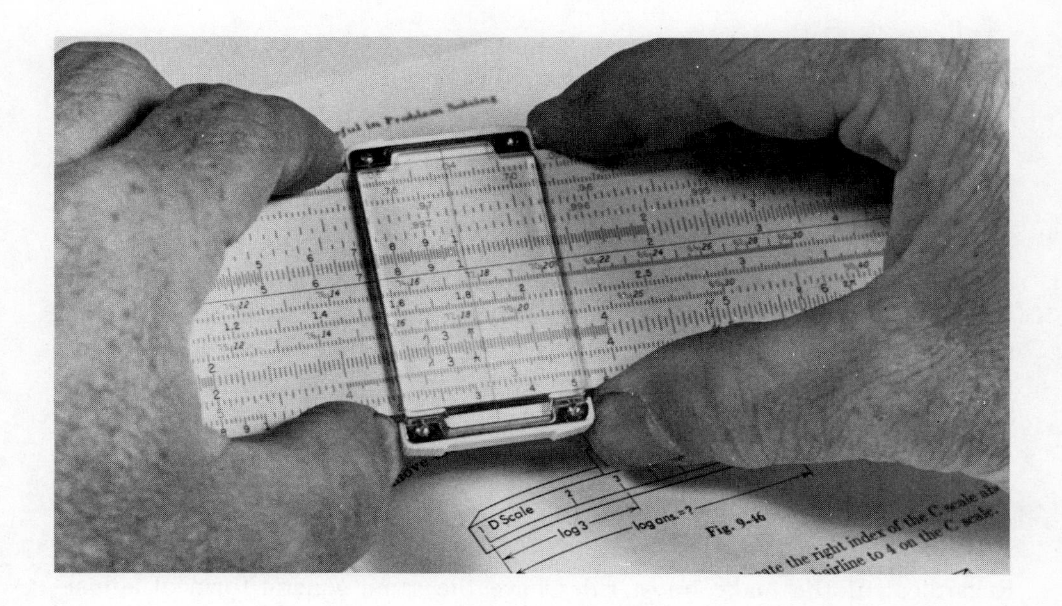

Illustration 2–1. In setting the indicator, a rolling motion with the forefingers will permit rapid and precise locations to be made. Keeping the fingers of both hands in contact with the indicator, exert slight forces toward each other with both hands.

Illustration 2–2. In moving the slide, use fingers to exert forces toward each other. A rolling motion with the forefinger aids in setting the indexes. Avoid pinching the frame because this will make the slide bind.

1. Settings usually can be made more rapidly by using two hands and holding the rule so that the thumbs are on the bottom with the backs of the hands toward the operator.
2. In moving either the indicator or the slide, the settings are easier to make if the index fingers and thumbs of both hands are used to apply forces toward each other than if only one hand is used to apply force. For example, in setting the indicator, put the forefinger of each hand against the respective edges of the indicator and move it by a combined squeezing and rolling motion of the forefingers. The same general procedure is used in setting the slide, where both hands exert forces toward each other. The student is cautioned in setting the slide not to squeeze the frame of the rule, since this will cause the slide to bind.

ADJUSTING THE RULE

Regardless of the make, most rules have the same general form of adjustment. The method of adjustment is simple but should not be applied in a hurry. It is desirable to use a magnifying glass, if one is available, to aid in lining up the scales and hairline.

To determine whether or not a rule needs adjustment, line up the indexes of the C and D scales. The indexes of the scales above and below the C and

D scales should also be aligned. If they do not coincide, slightly loosen the screws that clamp the top bar of the frame and carefully move the frame to the right or left until the indexes are aligned. Tighten the screws slightly and move the slide to check for proper friction. If the alignment and friction are satisfactory, tighten the frame screws to complete that part of the adjustment.

Next, test the hairline for proper alignment by setting the hairline over the indexes of the C and D scales and checking to see that the hairline also coincides with the other indexes on this side of the rule. If it does not coincide with all the scale indexes, slightly loosen the screws which hold the glass frame to the indicator. Rotate the frame slowly until the hairline coincides with the indexes on this side of the rule. Tighten the screws holding this frame; then, while the hairline is aligned on the indexes of the C and D scales, turn the rule over and check for the alignment of the hairline with the indexes of the scales on the other side of the rule. If the hairline does not coincide with the indexes on this side of the rule, loosen the screws on the indicator and make the necessary adjustment as before.

Check the tightness of all screws when the adjustment is completed. The student is cautioned not to use excessive force in tightening any screws, as the threads may become stripped. With reasonable care, a slide rule will usually require very little adjustment over a considerable period of time.

ACCURACY OF THE RULE

Most measurements made in scientific work contain from two to four significant figures; that is, digits which are considered to be reliable. Since the mathematical operations of multiplication, division, and processes involving roots and powers will not increase the number of significant figures when the answer is obtained, the slide rule maintains an accuracy of three or four significant figures. The reliability of the digits obtained from the rule depends upon the precision with which the operator makes his settings. It is generally assumed that with a 25-cm. slide rule, the error of the answer will not exceed about a tenth of 1 per cent. This is one part in a thousand.

A common tendency is to use more than three or four significant digits in such numbers as π (3.14159265 . . .) and ϵ (2.71828 . . .). The slide rule automatically "rounds off" such numbers to three or four significant figures, thus preventing false accuracy (such as can occur in longhand operations) from occurring in the answer.

In slide rule calculations the answer should be read to four significant figures if the first digit in the answer is 1 (10.62, 1.009, 1195., 1,833,000., etc.). In other cases the answer is usually read to three significant figures (2.95, 872., 54,600., etc.). The chance for error is increased as the number of operations in a problem increases. However, for average length operations, such as those required to solve the problems in this text, the fourth significant digit in the slide rule answer should not vary more than ±2 from the correct answer. Where only three significant digits are read from the rule, the third digit should be within ± 2 of the correct answer.

Example

+ 16.27, 16.26	within slide rule accuracy	within slide rule accuracy	0.0859, 0.0858 +
Correct Answer 16.25			**0.0857 Correct Answer**
− 16.24, 16.23	within slide rule accuracy	within slide rule accuracy	0.0856, 0.0855 −

Rules of modern manufacture are designed so that results read from the graduations are as reliable as the naked eye can distinguish. The use of magnifying devices may make the settings easier to locate but usually do not have an appreciable effect on the accuracy of the result.

INSTRUCTIONS FOR READING SCALE GRADUATIONS

Before studying the scales of the slide rule, let us review the reading of scale graduations in general. First let us examine a common 12-in. ruler (Figure 2–2).

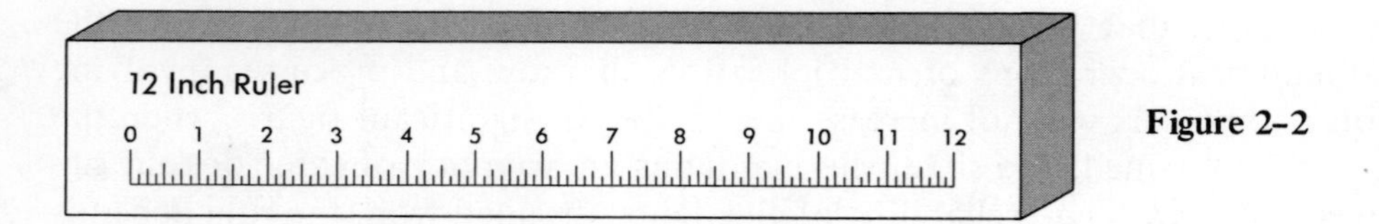

Figure 2–2

Example We see that the total length of 1 ft has been divided into 12 equal parts and that each part is further divided into quarters, eighths, and sixteenths. This subdivision is necessary so that the workman need not estimate fractional parts of an inch.

Example Measure the unknown lengths L_1 and L_2 as shown in Figure 2–3.

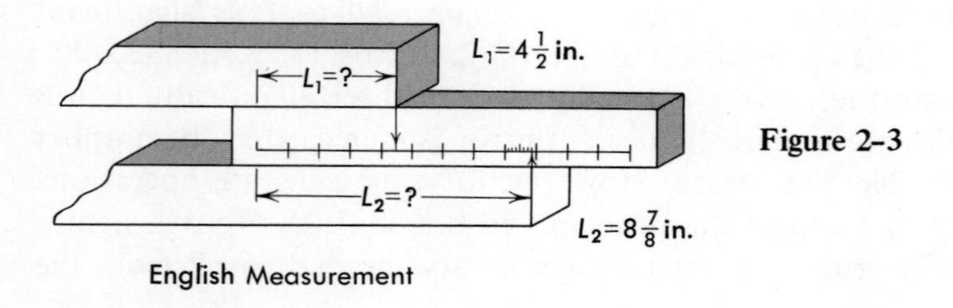

Figure 2–3

The English system of measurement as shown in Figure 2–3 is probably familiar to all students. The unit of length in the metric system which corresponds to the yard in the English system is called the *metre.* The meter is 39.37 in. in length. For convenience, the meter is divided into 100 equal parts called *centimetres,* and each centimeter is divided into ten equal parts called *millimetres.* Since we can express units and fractional parts of units as tenths or hundredths of the length of a unit, this system of measurement is preferred many times for engineering work.

Example Measure the unknown lengths L_1 and L_2 as shown in Figure 2–4.

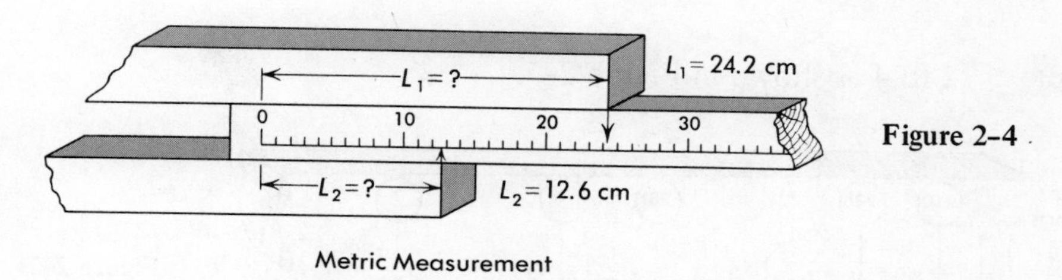

Figure 2–4

The scales of the slide rule are basically divided as in the metric system in that between each division there are ten subdivisions. However, the student will find that the main divisions are not equal distances apart. Sometimes the divisions will be subdivided by graduations, and at other times the student will need to estimate the subdivisions by eye. Let us examine the D scale of a slide rule (Figure 2–5).

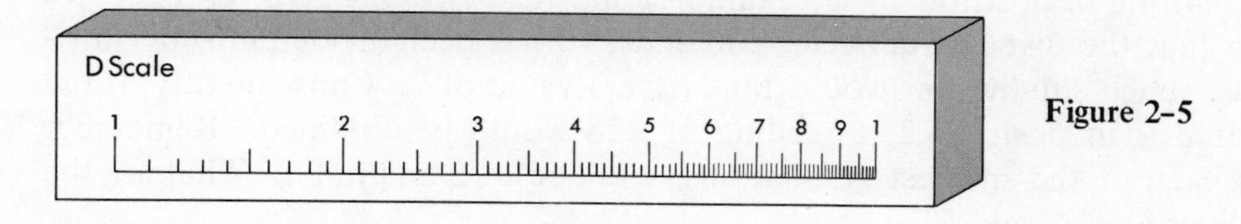

Figure 2–5

Since the graduations are so close together, let us examine the rule in three portions: from left index to 2, from 2 to 4, and from 4 to the right index.

Example Left index to 2 as shown in Figure 2–6.

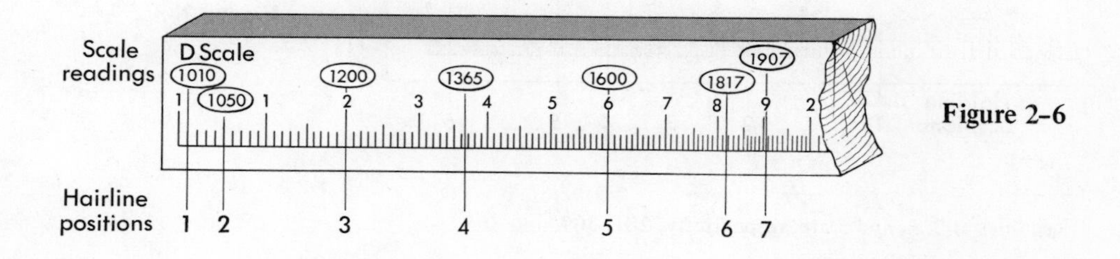

Figure 2–6

The student should refer to his own rule for comparisons as he studies the diagrams in this chapter. In the example using Figure 2–6, we note that from the left index (read as one-zero-zero) to the digit 1 (read as one-one-zero), there are ten graduations. The first is read as *one-zero-one* (101), the second as *one-zero-two* (102), and so on. Digit 2 is read as *one-two-zero* (120), digit 3 as *one-three-zero* (130), and so on. If need be, the student can subdivide by eye the distance between each of the small, unnumbered graduations. Thus, if the hairline is moved to position 4 (see example above), the reading would be *one-three-six-five* or 1365. Position 6 might be read as 1817 and position 7 as 1907. The student is reminded that each small graduation on this portion of the rule has a value of 1.

Example 2 to 4 as shown in Figure 2–7.

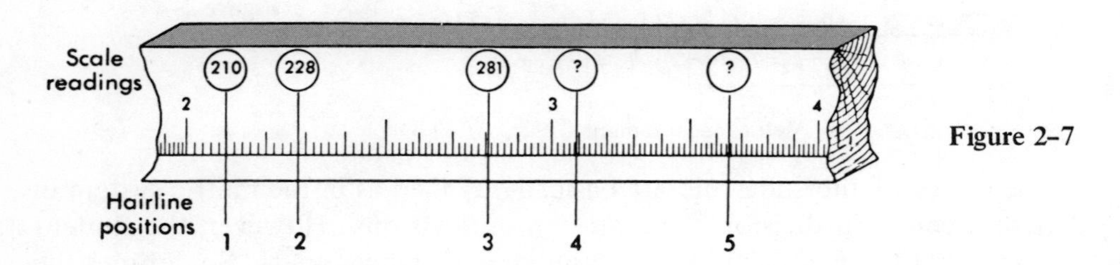

Figure 2–7

Since the distance between 2 and 3 is not as long as the distance from the left index to 2, no numbers are placed over the graduations. However, we can use the same reasoning and subdivide as in the previous examples. Set the hairline in position 1 (see example) and read *two-one-zero,* or 210. We note that the distance between 200 and 210 has been divided into five divisions. Each subdivision would thus have a value of 2. Consequently, if the hairline is in position 2, a reading of 228 would be obtained. Remember that each of the smallest graduations is valued at 2 and not 1. What are the readings at 3, 4, and 5?[1]

Example: 4 to the right index as shown in Figure 2–8.

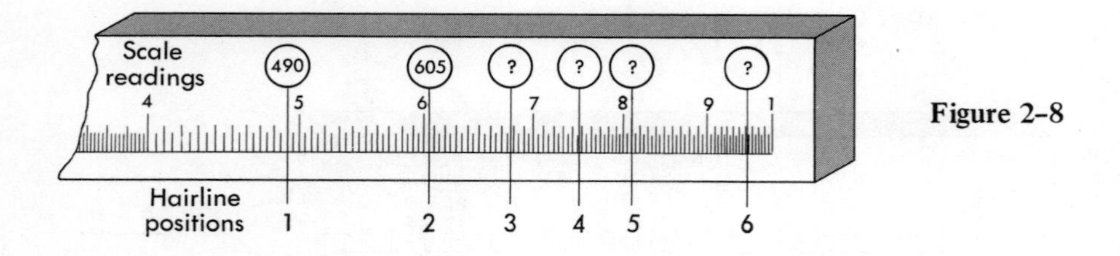

Figure 2–8

[1]Readings at 3, 4, and 5 are, respectively, 281, 309, and 365.

The distance between 4 and 5 is still shorter than the distance between 3 and 4, and it becomes increasingly more difficult to print such small subdivisions. For this reason there are ten main divisions between 4 and 5, each of which is subdivided into two parts. With this type of marking it is possible to read two figures and estimate the third, or to get three significant figures on all readings. If the hairline is set as indicated in position 1, the reading would be *four-nine-zero* (490), and position 2 would give *six-zero-five* (605). What are the readings at hairline positions 3, 4, 5, and 6?[2]

Problems on scale readings

Set hairline to	*Read answer on*								
	ST scale	*T scale*	*LL_3 scale*	*CI scale*	*K scale*	*DF scale*	*LL_{01} scale*	*LL_2 scale*	*L scale*
1. 210 on D									
2. 398 on D									
3. 1056 on D									
4. 1004 on D									
5. 866 on D									
6. 222 on D									
7. 1196 on D									
8. 439 on D									
9. 5775 on D									
10. 2325 on D									
11. 917 on D									
12. 323 on D									
13. 1077 on D									
14. 1854 on D									
15. 268 on D									
16. 833 on D									
17. 551 on D									
18. 667 on D									
19. 8125 on D									
20. 406 on D									
21. 918 on D									
22. 805 on D									
23. 1466 on D									
24. 288 on D									
25. 466 on D									
26. 798 on D									
27. 1107 on D									
28. 396 on D									
29. 1999 on D									
30. 998 on D									

[2] Readings at 3, 4, 5, and 6 are, respectively, 678, 746, 810, and 963.

If the student has followed the reasoning thus far, he should have little trouble in determining how to read an indicated value on any scale of the slide rule. Several of the problems on page 21 should be worked, and the student should thoroughly understand the principle of graduation subdivision before he attempts to delve further into the uses of the slide rule.

It is suggested that one have a good understanding of logarithms before proceeding to learn the operational aspects of the slide rule. Those who may desire to review these principles should refer to Appendix I.

CONSTRUCTION OF THE SCALES

Let us examine how the main scales (C and D) of the rule are constructed. As a basis for this examination, let us set up a scale of some length with a beginning graduation called a *left index* and an end graduation called a *right index* as in Figure 2–9.

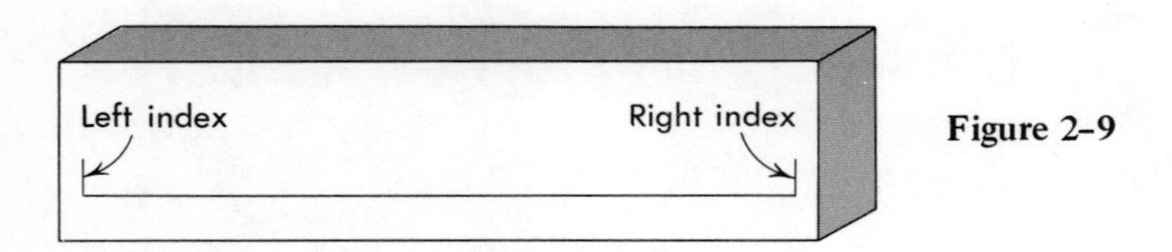

Figure 2–9

Next let us subdivide this scale into ten equal divisions and then further subdivide each large division into ten smaller divisions as shown in Figure 2–10. We call this the *L scale.*

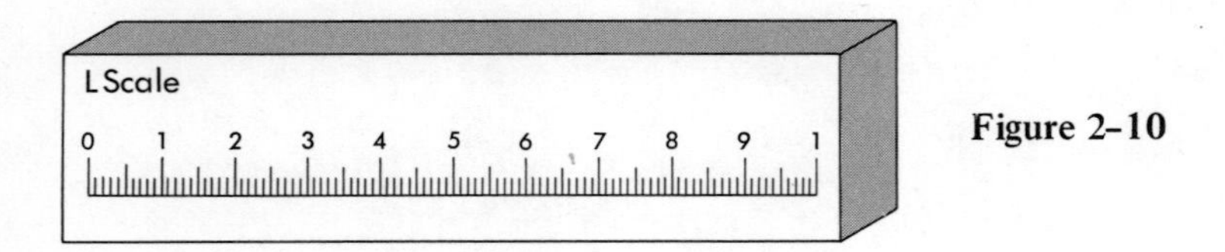

Figure 2–10

Let us place a blank scale beneath this L scale so that the left index of the L scale will coincide with the left index of the blank scale as shown in Figure 2–11. We shall call the blank scale the *D scale.*

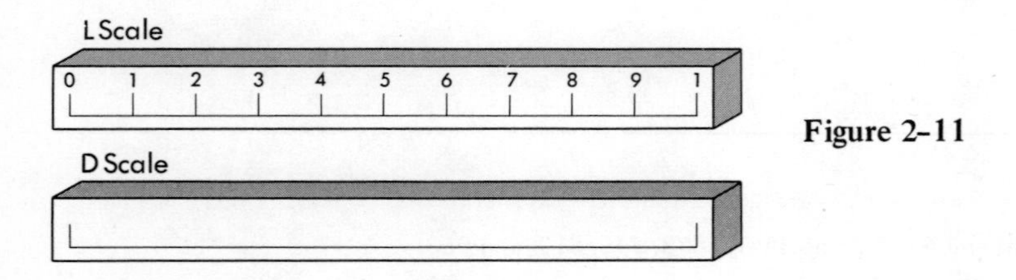

Figure 2–11

Now let us graduate the D scale in such a way that each division mark is directly beneath the mark on the L scale that represents the mantissa of the logarithm of the number. Before examining the scales closer, we should note that the mantissa of 2 is 0.3010, the mantissa of 3 is 0.4771, the mantissa of 4 is 0.6021, and the mantissa of 5 is 0.6990 as shown in Figure 2–12 (see also page 223–224).

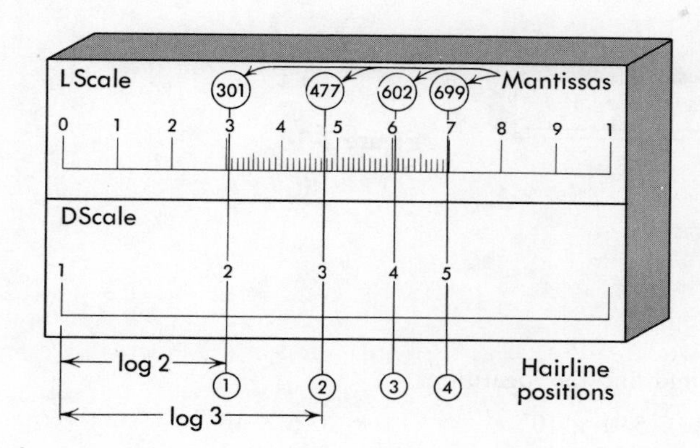

Figure 2–12

If the student will examine his rule, he will find a C or D scale and an L scale. The C and D scales are identical, so use the D scale since it is printed on the body of the rule. Several problems should be worked, determining the logarithms of numbers by using the slide rule.

Remember to:

1. Set the number on the D scale.
2. Read the mantissa of the number on the L scale.
3. Supply the characteristic, using the *characteristic rules* given in the discussion on logarithms in Appendix I.

Example What is the logarithm of 55.8? Use Figure 2–13.

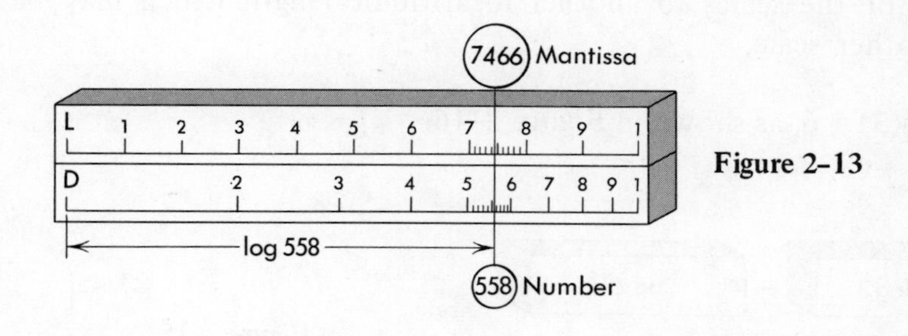

Figure 2–13

From slide rule:	Mantissa of 55.8 = 0.7466
From characteristic rules:	Characteristic of 55.8 = 1.0000
Therefore	log of 55.8 = **1.7466**

From the preceding example, we can see that the D scale is so constructed that each number lies below the mantissa of its logarithm. Also we note that the distance from the left index of the D scale to any number on the D scale represents (in length) the mantissa of the number as shown in Figure 2–14. Since the characteristic of a logarithm is governed merely by the location of the decimal point, we can delay its determination for the time being.

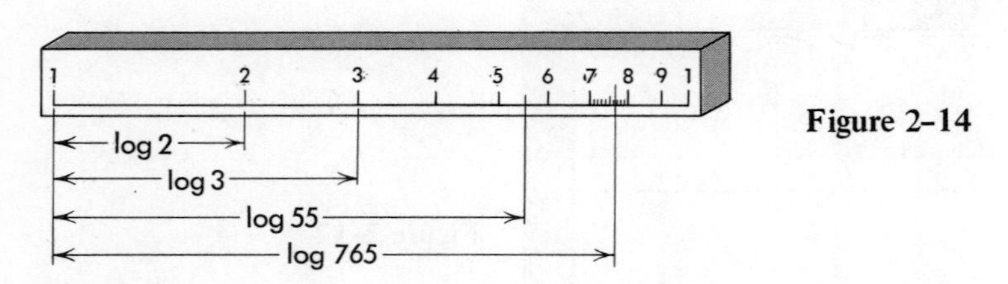

Figure 2–14

Problems

2-1. Use the slide rule and find the logarithms.

a. 894.
b. 1.845
c. 0.438
d. 81.5
e. 604.
f. 7.41
g. 11.91
h. 215.
i. 993,000.
j. 5.91×10^7
k. 9.06×10^{-4}
l. 66.9×10^8
m. 155.8×10^2
n. 23.66×10^{-4}
o. 0.06641×10^8
p. 9.33×10^{-2}
q. 29.8×10^{-1}
r. 0.552×10^6
s. 33.6×10^{-9}
t. 4.40×10^3
u. 98,700
v. 40.3×10^{-9}
w. 21.8×10^9
x. 1.057×10^{-3}
y. $719. \times 10^5$
z. 49.2×10^7

MULTIPLICATION

As shown in Figure 2–15, the C and D scales are divided logarithmically with all graduations being marked with their corresponding antilogarithms. These scales can be used for multiplication by adding a given logarithmic length on one of the scales to another logarithmic length which may be found on the other scale.

Example (2)(3) = 6, as shown in Figure 2–15.

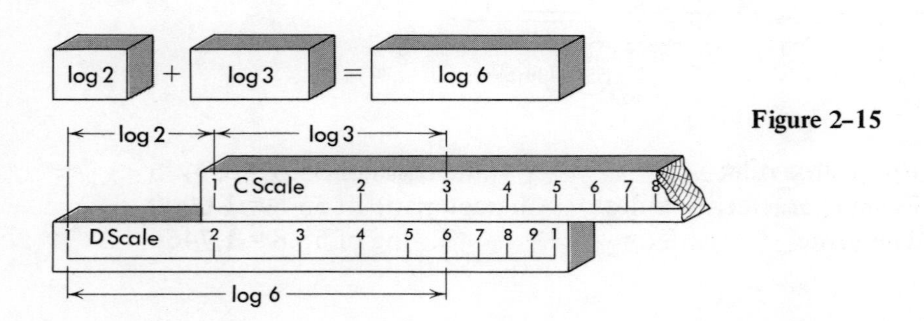

Figure 2–15

Procedure

1. Set the left index of the C scale above the digit 2 on the D scale.
2. Move the hairline to the right until it is directly over 3 on the C scale.
3. Read the answer (6) directly under the hairline on the D scale.

The A and B scales are also divided logarithmically, but their overall lengths are only one half the lengths of the C and D scales. Therefore, although the A and B scales can also be used for multiplication and division, their shortened lengths will diminish the accuracy of the readings.

Similarly other pairs of scales of the slide rule may be used to perform multiplication if they are graduated logarithmically. A majority of slide rules have at least one set of folded scales that can be used for this purpose. Most frequently they are folded at π (3.14159 . . .). Special use of these scales will be explained later in this chapter.

In some cases, when the logarithm of one number is added to the logarithm of another number, the multiplier extends out into space, and it is impossible to move the indicator to the product (Figure 2–16).

Example (3)(4) = ?, as shown in Figure 2–16).

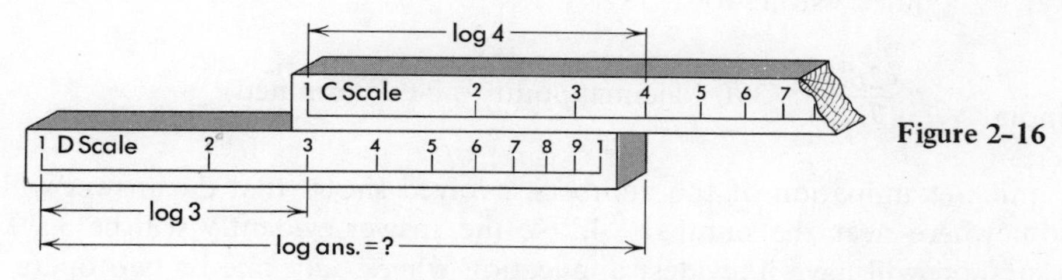

Figure 2–16

In this case it is necessary to relocate the right index of the C scale above the figure 3 on the D scale and move the hairline to 4 on the C scale as shown in Figure 2–17.

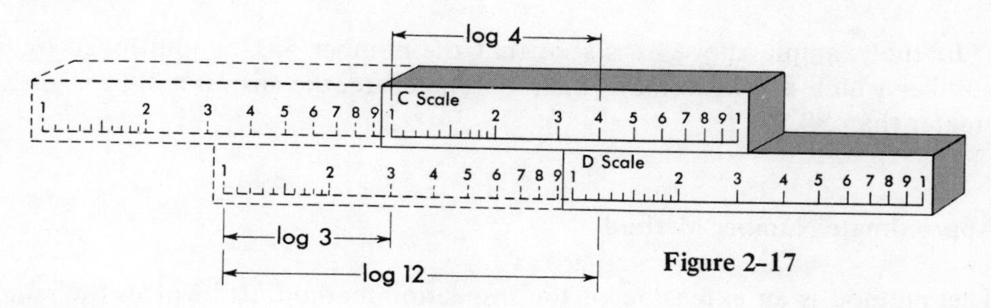

Figure 2–17

Procedure

1. Set the right index of the C scale above the digit 3 on the D scale.
2. Move the hairline to the left until it is directly over 4 on the C scale.
3. Read answer (12) directly under the hairline on the D scale.

The location of the decimal point in multiplication problems is ascertained either by inspection or by applying one of the several methods explained in the following paragraphs.

METHODS OF DETERMINING DECIMAL POINT LOCATION

Several methods which may be used are given below. Although these methods by no means include all ways to determine the decimal point location, they will be suitable for instruction of students, particularly those having an elementary mathematical background.

Inspection Method

This is the simplest method and consists of determining the decimal point location by observing the location of the decimal point in the numbers involved in a slide rule operation and locating the decimal point in the answer by a quick estimation.

Example $\frac{(28.1)}{(7.20)} = \mathbf{390}$ (decimal point to be determined)

A quick examination of the numbers involved shows that the answer will be somewhere near the number "4," so the answer evidently will be 3.90. This method will have its widest application where only one or two operations are involved and where the numbers lie between 1 and 100.

Example $(1.22)(58.2) = \mathbf{71.0}$

In the example above, it is seen that the number 58.2 is multiplied by a number which is a little more than 1. Therefore, the answer will be slightly greater than 58.2.

Approximate Number Method

This method is an extension of the inspection method. It involves the same general procedures except that the numbers used in a problem are "rounded off" and written down and an approximate answer is obtained that will show the decimal point location.

Example $(37.6)(0.188)(5.71)(11.92) = \mathbf{481}$ (decimal point to be located)

Rewrite, using simple numbers that are near in value to the problem numbers.

$$[(40)(0.2)]\,[(6)(10)] = (8)(60) = \mathbf{480}$$

This shows that the answer in the example problem should be expressed as 481.

A problem that is more involved can be solved by this method, as shown by the following example.

Example $\dfrac{(12{,}560)(0.0387)}{(594{,}000)} = \mathbf{818}$ (decimal point to be determined)

Using simple numbers near in value to the problem numbers, write the same problem:

$$\frac{(12{,}000)(0.04)}{(600{,}000)} = \mathbf{0.0008}$$

By cancellation the numbers can be simplified still further to obtain an approximate answer of 0.0008. One way of doing this would be to divide 12,000 into 600,000, obtaining a value of 50 in the denominator. This value of 50 divided into 0.04 gives 0.0008. Referring to the original problem, the decimal point must be located to give an answer of 0.000819.

Scientific notation or power-of-ten method

The power-of-ten or scientific notation method is a variation of the characteristic method discussed on page 29. In this method the numbers in the problem are expressed as a single digit, a decimal point, the remaining digits, and followed by the number "10" raised to the appropriate power. This process simplifies the numbers, and the decimal point in the answer can be determined by inspection or by the approximation method. For a review of scientific notation refer to page 9.

Example

$$(15.9)(0.0077)(30500)(4660) = \mathbf{1740} \text{ (decimal point to be located)}$$

Write the same problem with each number expressed as a digit, decimal point, and the remaining digits followed by the appropriate power of 10.

$$(1.59 \times 10^{1})(7.7 \times 10^{-3})(3.05 \times 10^{4})(4.66 \times 10^{3}) = \mathbf{174.0 \times 10^{5}}$$

Since all the numbers are now expressed as numbers between 1 and 10, followed by 10 to a power, the approximate value of the multiplication can be determined rapidly, by inspection, to be about 170. The power of 10 is obtained by adding algebraically the powers of 10 of each of the rewritten numbers. The answer to the original problem is therefore 174.0×10^5, or 17,400,000, or 1.740×10^7.

Example $\quad \dfrac{(18{,}500)(307)}{(0.552)} = \mathbf{1585}$ (decimal point to be located)

Rewrite the problem using powers of 10:

$$\frac{(1.85 \times 10^4)(3.07 \times 10^2)}{(5.52 \times 10^{-1}} = \mathbf{1.585 \times 10^7}$$

By inspection and approximation the product of the numerator will be found to be near 9, and dividing 5.52 into it will give about 1.6. This procedure determines the decimal point location for the digits of the answer. The powers of 10 are added algebraically to give 10^7, which completes the decimal point location in the answer. The answer may be rewritten as 15,850,000 if desired.

Digit method

In this method the numbers of digits in each number are counted and the following rules apply.

Multiplication Add the number of digits to the left of the decimal of each number to be multiplied. This will give the number of digits to the left of the decimal in the answer. If the slide projects to the right, subtract 1 from the number of digits to be pointed off.

Example $\quad (27{,}300)(15.1) = \mathbf{412{,}000}$

There are five digits to be counted in the first number and only two digits in the second number. Since the slide projects to the right, subtract 1. There will be six digits to the left of the decimal point in the answer.

Division Subtract the number of digits to the left of the decimal in the denominator from the number of digits to the left of the decimal in the numerator to obtain the number of digits to the left of the decimal in the answer. If the slide projects to the right in division, add one digit more to be pointed off.

Example $\quad \dfrac{(12.88)}{(466)} = \mathbf{0.0276}$

Subtracting three digits in the denominator from two digits in the numerator gives (–1) digit to be located in the answer. Inspection shows that the answer will be a decimal quantity. In any case where decimal numbers are encountered, the method of counting the digits is to begin at the decimal point and count the number of zeroes between the decimal point and the first digit that is not zero to the right of the decimal. Since the digit difference shown above is (−1), there must be one zero between the decimal point and the first significant figure, which gives an answer of 0.0276. The student will observe that the digit count of decimal numbers is considered as a minus quantity and that the addition and subtraction of the digit count must take into account any minus signs.

Variations and extensions of these methods may readily be set up to solve problems involving roots and powers. Some schools prefer the "characteristic" or "projections method" to determine decimal point location, and this method is given in detail in the discussions which follow.

Characteristic method

Projection rule for multiplication This method of decimal point location is recommended for students who are inexperienced in slide rule computations:

1. Before attempting to solve the problem, place the characteristic of each quantity above or below it.
2. Solve for the sum of the characteristics by simple addition, and place this number above the space for the answer.
3. Begin the multiplication with the slide rule, and each time the left index of the C scale extends past the left index of the D scale, add a (+1) to the sum of the characteristics previously determined.
4. Add the original sum to the +1's obtained from left extensions. The total number is the characteristic of the answer.

Example

one left extension
↓

Characteristics (0) + (0) → (0) + 1 = +1 ← characteristic of answer

(5) (3) = 15 *Answer*

ESTIMATION OF ANSWER BY SCIENTIFIC NOTATION:

$(5)(3) = \mathbf{1.5(10)^1}$ ← ESTIMATED ANSWER

Example

one left extension
↓

Characteristics (+2) + (−3) → (−1) + 1 = 0

(390) (0.0030) = 1.17 *Answer*

ESTIMATION OF ANSWER BY SCIENTIFIC NOTATION:

$(4)(10)^2(3)(10)^{-3} = \mathbf{1.2(10)^0}$ ← ESTIMATED ANSWER

Example

								two left ↓ extensions	
Characteristics	(−3)	+	(+1)	+	(+2)	+	(+4)	→ (+4) + 2 = +6	
	(0.001633)		(79.1)		(144)		(96,500)	= **1,800,000**	*Answer*

> ESTIMATION OF ANSWER BY SCIENTIFIC NOTATION:
> $(2)(10)^{-3}(8)(10)^1(1)(10)^2(10)^5 = \mathbf{1.6(10)^6}$ ← ESTIMATED ANSWER

Example

								three left ↓ extensions	
Characteristics	(+1)	+	(+3)	+	(−3)	+	(−4)	→ (−3) + 3 = 0	
	(73.7)		(4460)		(0.00704)		(0.000853)	= **1.975**	*Answer*

> ESTIMATION OF ANSWER BY SCIENTIFIC NOTATION:
> $(7)(10)^1(4)(10)^3(7)(10)^{-3}(9)(10)^{-4} = \mathbf{1.8(10)^0}$ ← ESTIMATED ANSWER

Example

								two left extensions ↓	
Characteristics	(+2)	+	(+2)	+	(0)	→	(+4) +	1 + 1 = +6	
	(861)		(204)		(9.0)		= 1,580,000 or $(1.58)(10)^6$		*Answer*

> ESTIMATION OF ANSWER BY SCIENTIFIC NOTATION:
> $(9)(10)^2(2)(10)^2(9) = \mathbf{1.6(10)^6}$ ← ESTIMATED ANSWER

Multiplication practice problems

2-2. $(23.8)(31.6) = \mathbf{(7.52)(10)^2}$
2-3. $(105.6)(4.09) = \mathbf{(4.32)(10)^2}$
2-4. $(286{,}000)(0.311) = \mathbf{(8.89)(10)^4}$
2-5. $(0.0886)(196.2) = \mathbf{(1.738)(10)^1}$
2-6. $(0.769)(47.2) = \mathbf{(3.63)(10)^1}$
2-7. $(60.7)(17.44) = \mathbf{(1.059)(10)^3}$
2-8. $(9.16)(115.7) = \mathbf{(1.06)(10)^3}$
2-9. $(592.)(80.1) = \mathbf{(4.74)(10)^4}$
2-10. $(7.69 \times 10^3)(0.722 \times 10^{-6}) = \mathbf{(5.55)(10)^{-3}}$
2-11. $(37.5 \times 10^{-1})(0.0974 \times 10^{-3}) = \mathbf{(3.65)(10)^{-4}}$
2-12. $(23.9)(0.715)(106.2) = \mathbf{(1.815)(10)^3}$
2-13. $(60.7)(1059)(237{,}000) = \mathbf{(1.523)(10)^{10}}$
2-14. $(988)(8180)(0.206) = \mathbf{(1.665)(10)^6}$
2-15. $(11.14)(0.0556)(76.3 \times 10^{-6}) = \mathbf{(4.73)(10)^{-5}}$
2-16. $(72.1)(\pi)(66.1) = \mathbf{(1.497)(10)^4}$
2-17. $(0.0519)(16.21)(1.085) = \mathbf{(9.13)(10)^{-1}}$
2-18. $(0.001093)(27.6)(56{,}700) = \mathbf{(1.710)(10)^3}$
2-19. $(0.379)(0.00507)(0.414) = \mathbf{(7.96)(10)^{-4}}$
2-20. $(16.05)(23.9)(0.821) = \mathbf{(3.15)(10)^2}$
2-21. $(1009)(0.226)(774) = \mathbf{(1.765)(10)^5}$

2-22. $(316)(825)(67{,}600) = \mathbf{(1.762)(10)^{10}}$
2-23. $(21{,}000)(0.822)(16.92) = \mathbf{(2.92)(10)^{5}}$
2-24. $(0.707)(80.6)(0.451) = \mathbf{(2.57)(10)^{1}}$
2-25. $(1.555 \times 10^3)(27.9 \times 10^5)(0.902 \times 10^{-7}) = \mathbf{(3.91)(10)^{2}}$
2-26. $(0.729)(10)^3(22{,}500)(33.2) = \mathbf{(5.45)(10)^{8}}$
2-27. $(18.97)(0.216)(899)(\pi)(91.2) = \mathbf{(1.055)(10)^{6}}$
2-28. $(7160)(0.000333)(26)(19.6)(5.01) = \mathbf{(6.09)(10)^{3}}$
2-29. $(1.712)(89{,}400)(19.5)(10^{-5})(82.1) = \mathbf{(2.45)(10)^{3}}$
2-30. $(62.7)(0.537)(0.1137)(0.806)(15.09) = \mathbf{(4.66)(10)^{1}}$
2-31. $(10)^6(159.2)(144)(7{,}920{,}000)(\pi) = \mathbf{(5.70)(10)^{17}}$
2-32. $(0.0771)(19.66)(219)(0.993)(7.05) = \mathbf{(2.32)(10)^{3}}$
2-33. $(15.06)(\pi)(625)(0.0963)(43.4) = \mathbf{(1.236)(10)^{5}}$
2-34. $(2160)(1802)(\pi)(292)(0.0443) = \mathbf{(1.582)(10)^{8}}$
2-35. $(437)(1.075)(0.881)(43{,}300)(17.22) = \mathbf{(3.09)(10)^{8}}$
2-36. $(\pi)(91.6)(555)(0.673)(0.00315)(27.7) = \mathbf{(9.38)(10)^{3}}$
2-37. $(18.01)(22.3)(1.066)(19.36)(10)^{-5} = \mathbf{(8.29)(10)^{-2}}$
2-38. $(84.2)(15.62)(921)(0.662)(0.1509) = \mathbf{(1.210)(10)^{5}}$
2-39. $(66{,}000)(25.9)(10.62)(28.4)(77.6) = \mathbf{(4.00)(10)^{10}}$
2-40. $(55.1)(7.33 \times 10^{-8})(76.3)(10)^5(0.00905) = \mathbf{(2.79)(10)^{-1}}$
2-41. $(18.91)(0.257)(0.0811)(92{,}500)(\pi) = \mathbf{(1.145)(10)^{5}}$

Multiplication problems

2-42. (46.8)(11.97)
2-43. (479.)(11.07)
2-44. (9.35)(77.8)
2-45. (10.09)(843,000.)
2-46. (77,900)(0.467)
2-47. (123.9)(0.00556)
2-48. (214.9)(66.06)
2-49. (112.2)(0.953)
2-50. (87.0)(1.006)
2-51. (1,097,000)(1.984)
2-52. (43.8)(0.000779)
2-53. (31.05)(134.9)
2-54. (117.9)(98.9)
2-55. (55.6)(68.1)
2-56. (1.055)(85.3)
2-57. (33,050.)(16,900.)
2-58. (6.089)(44.87)
2-59. (34.8)(89.7)
2-60. (43,900.)(19.07)
2-61. (41.3)(87.9)
2-62. (99.7)(434,000.)
2-63. (0.0969)(0.1034)(0.1111)(0.1066)
2-64. $(1.084 \times 10^{-5})(0.1758 \times 10^{13})(66.4)(0.901)$
2-65. $(234.5)(10)^4(21.21)(0.874)(0.0100)$
2-66. $(\pi)(26.88)(0.1682)(0.1463)(45.2)(1.007)$
2-67. $(75.8)(0.1044 \times 10^8)(10)^{-2}(54{,}000)(0.769)$
2-68. $(34.5)(31.09)(10)^{-6}(54.7)(0.677)(0.1003)$
2-69. $(6.08)(5.77)(46.8)(89.9)(3.02)(0.443)(\pi)$
2-70. $(1.055)(6.91)(31.9)(11.21)(\pi)(35.9)(4.09)$

2-71. (10.68)(21.87)
2-72. (88,900.)(54.7)
2-73. (113,900.)(48.1)
2-74. (95,500.)(0.000479)
2-75. (0.0956)(147.2)(0.0778)
2-76. (15.47)(82.5)(975,000.)
2-77. (37.8)(22,490,000.)(0.15)
2-78. (1.048)(0.753)(0.933)
2-79 $(1.856)(10)^3(21.98)$
2-80. (57.7)(46.8)(3.08)
2-81. (0.045)(0.512)(115.4)
2-82. (0.307)(46.3)(7.94)
2-83. $(2.229)(86.05)(16{,}090.)(\pi)$
2-84. (44,090.)(38.9)(667.)(55.9)
2-85. (568.)(46.07)(3.41)(67.9)
2-86. (75.88)(0.0743)(0.1185)(0.429)
2-87. $(10)^{-7}(69.8)(11.03)(0.901)$
2-88. $(46.3)(0.865)(10)^{-9}(0.953)(\pi)$
2-89. (665.)(35,090)(0.1196)(0.469)
2-90. (888.)(35.9)(77.9)(0.652)
2-91. $(43.4)(0.898)(70.09)(0.113)(\pi)$

DIVISION

Multiplication is merely the process of mechanically adding the logarithms of the quantities involved. From a review of the principles of logarithms, it follows that division is merely the process of mechanically subtracting the logarithm of the divisor from the logarithm of the dividend.

Example $\frac{(8)}{(2)} = 4$, as shown in Figure 2–18.

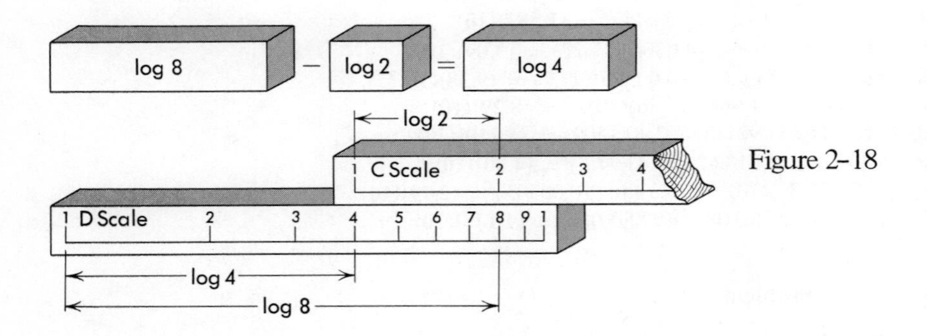

Figure 2–18

Procedure

1. Set the divisor (2) on the C scale directly above the dividend (8), which is located on the D scale.
2. Read the answer (4) on the D scale directly under the left index of the C scale.

For location of the decimal point in division problems it is suggested that the "scientific notation" method be used (described on page 27). As an alternate method the following *Projection Rule* may be followed.

Projection rule for division

1. Locate the characteristic of the dividend above it and the characteristic of the divisor below it.
2. Subtract the characteristic of the divisor from the characteristic of the dividend.
3. For every left extension of the C scale's left index, add a (–1) to the total characteristic already obtained.
4. The sum is the characteristic of the answer.

Example

left extension ↓ characteristic of answer ↓

$$\frac{\overset{(+2)}{(575)}}{\underset{(0)}{(6.05)}} = (9.50)(10)^1 \qquad (+2) - (0) \rightarrow +2 - 1 = +1$$

ESTIMATION OF ANSWER BY SCIENTIFIC NOTATION:

$$\frac{6(10)^2}{6} = 1(10)^2 \leftarrow \text{ESTIMATED ANSWER}$$

Example

left extension ↓ characteristic of answer ↓

$$\frac{\overset{(-1)}{(0.465)}}{\underset{(+1)}{(54)}} = (8.61)(10)^{-3} \qquad (-1) - (+1) \rightarrow -2 - 1 = -3$$

ESTIMATION OF ANSWER BY SCIENTIFIC NOTATION:

$$\frac{5(10)^{-1}}{5(10)^1} = 1(10)^{-2} \leftarrow \text{ESTIMATED ANSWER}$$

Division practice problems

2-92. $(29.6) \div (18.02) = \mathbf{1.641}$
2-93. $(1.532) \div (72.6) = \mathbf{(2.11)(10)^{-2}}$
2-94. $(0.1153) \div (70.3) = \mathbf{(1.64)(10)^{-3}}$
2-95. $(89.3) \div (115.6) = \mathbf{(7.72)(10)^{-1}}$
2-96. $(0.1052) \div (33.6) = \mathbf{(3.13)(10)^{-3}}$
2-97. $(40.2) \div (50.8) = \mathbf{(7.91)(10)^{-1}}$
2-98. $(0.661) \div (70{,}500) = \mathbf{(9.38)(10)^{-6}}$
2-99. $(182.9) \div (0.00552) = \mathbf{(3.31)(10)^{4}}$
2-100. $(0.714) \div (98{,}200) = \mathbf{(7.27)(10)^{-6}}$
2-101. $(4.36) \div (80{,}300) = \mathbf{(5.43)(10)^{-5}}$
2-102. $(1.339) \div (22.6 \times 10^4) = \mathbf{(5.92)(10)^{-6}}$
2-103. $(17.03) \div (76.3) = \mathbf{(2.23)(10)^{-1}}$
2-104. $(0.511) \div (0.281) = \mathbf{1.819}$
2-105. $(67.7) \div (91{,}300) = \mathbf{(7.42)(10)^{-4}}$
2-106. $(5.04) \div (29{,}800) = \mathbf{(1.691)(10)^{-4}}$
2-107. $(18.35) \div (0.921) = \mathbf{(1.992)(10)^{1}}$
2-108. $(29.6 \times 10^5) \div (0.905) = \mathbf{(3.27)(10)^{6}}$
2-109. $(0.1037) \div (92.5 \times 10^5) = \mathbf{(1.121)(10)^{-8}}$
2-110. $(537) \div (15.63 \times 10^{-7}) = \mathbf{(3.44)(10)^{8}}$
2-111. $(26{,}300) \div (84.3 \times 10^5) = \mathbf{(3.12)(10)^{-3}}$
2-112. $(6.370) \div (0.733) = \mathbf{(8.69)(10)^{3}}$
2-113. $(1.066) \div (7.51 \times 10^3) = \mathbf{(1.419)(10)^{-4}}$
2-114. $(29.6 \times 10^4) \div (0.973) = \mathbf{(3.04)(10)^{5}}$
2-115. $(0.912) \div (10.31 \times 10^{\ 5}) = \mathbf{(8.85)(10)^{3}}$

2-116. $(17.37 \times 10^{-4}) \div (0.662) = \mathbf{(2.62)(10)^{-3}}$
2-117. $(0.693 \times 10^{5}) \div (1.008 \times 10^{-6}) = \mathbf{(6.88)(10)^{10}}$
2-118. $(89.1 \times 10^{3}) \div (189.3 \times 10^{4}) = \mathbf{(4.71)(10)^{-2}}$
2-119. $(0.617) \div (29{,}600) = \mathbf{(2.08)(10)^{-5}}$
2-120. $(18.06 \times 10^{7}) \div (15.29) = \mathbf{(1.181)(10)^{7}}$
2-121. $(56.8)(10)^{4} \div (29.6)(10)^{-3} = \mathbf{(1.919)(10)^{7}}$
2-122. $(183{,}600) \div (76.3 \times 10^{-3}) = \mathbf{(2.41)(10)^{6}}$
2-123. $(75.9 \div (0.000813) = \mathbf{(9.34)(10)^{4}}$
2-124. $(43.6) \div (0.0837) = \mathbf{(5.21)(10)^{2}}$
2-125. $(156.8 \times 10^{3}) \div (0.715) = \mathbf{(2.19)(10)^{5}}$
2-126. $(216 \times 10^{-3}) \div (1557) = \mathbf{(1.387)(10)^{-4}}$
2-127. $(88.3 \times 10^{-1}) \div (29.1 \times 10^{-4}) = \mathbf{(3.03)(10)^{3}}$
2-128. $(1.034 \times 10^{3}) \div (0.706 \times 10^{-8}) = \mathbf{(1.465)(10)^{11}}$
2-129. $(55.2)(10)^{3} \div (0.1556 \times 10^{3}) = \mathbf{(3.55)(10)^{2}}$
2-130. $(0.01339) \div (1896 \times 10^{5}) = \mathbf{(7.06)(10)^{-11}}$
2-131. $(4{,}030 \times 10^{-7}) \div (75.3 \times 10^{-9}) = \mathbf{(5.35)(10)^{3}}$

Problems in division

2-132. $\frac{89.9}{45.}$

2-133. $\frac{147.}{22.}$

2-134. $\frac{9.06}{7.1}$

2-135. $\frac{1{,}985.}{78.55}$

2-136. $\frac{19{,}230.}{64.88}$

2-137. $\frac{87{,}600.}{43.8}$

2-138. $\frac{54.8}{9.10}$

2-139. $\frac{0.877}{33.07}$

2-140. $\frac{11.44}{24.9}$

2-141. $\frac{187{,}900.}{71.45}$

2-142. $\frac{0.00882}{87.04}$

2-143. $\frac{0.675}{54.8}$

2-144. $\frac{87.9}{45.7}$

2-145. $\frac{164{,}800.}{3.88}$

2-146. $\frac{7.09 \times 10^{3}}{18.45}$

2-147. $\frac{(0.001755)}{(6.175)}$

2-148. $\frac{(0.0000559)}{(0.00659)}$

2-149. $\frac{(5.065)}{(0.0003375)}$

2-150. $\frac{(469{,}000)}{(793)}$

2-151. $\frac{(5{,}100{,}000)}{(933 \times 10^{5})}$

2-152. $\frac{(3765 \times 10^{3})}{(760.3)}$

2-153. $\frac{(4917)}{(0.391)}$

2-154. $\frac{(5516)}{(1.65)}$

2-155. $\frac{(0.0916)}{(0.331)}$

2-156. $\frac{(193.7)}{5.06}$

2-157. $\frac{(113.05)}{(72.35)}$

2-158. $\frac{(32.33)}{(46.77)}$

2-159. $\frac{(3.17)}{(3.1416)}$

2-160. $\frac{(0.221)}{(56.91)}$

2-161. $\frac{(233.17)}{(5506)}$

2-162. $\frac{(72.13)}{(52.03)}$

2-163. $\frac{(6607)}{(1.91 \times 10^{5})}$

2-164. $\frac{(1.993 \times 10^{-8})}{(72.31 \times 10^{-6})}$

2-165. $\frac{(461 \times 10^{3})}{(0.003617)}$

2-166. $\frac{(9903 \times 10^{-5})}{(47.31 \times 10^{3})}$

2-167. $\frac{0.711}{11{,}980.}$

2-168. $\frac{0.01253}{66.8}$

2-169. $\frac{0.974}{1.058}$

2-170. $\frac{0.000497}{38.9 \times 10^{-5}}$

2-171. $\frac{48.6 \times 10^{-9}}{1.977 \times 10^{5}}$

2-172. $\frac{69{,}990. \times 10^{18}}{43.9 \times 10^{-2}}$

2-173. $\frac{5.06 \times 10^{-7}}{0.001853 \times 10^{9}}$

2-174. $\frac{1.097 \times 10^{-6}}{458. \times 10^{-1}}$

2-175. $\frac{89.99 \times 10^{-3}}{40.7 \times 10^{-6}}$

2-176. $\dfrac{659{,}000}{0.1148 \times 10^{-3}}$

2-177. $\dfrac{883.8}{3.89 \times 10^{-11}}$

2-178. $\dfrac{15.06 \times 10^{-7}}{33.8 \times 10^{-1}}$

2-179. $\dfrac{1.095}{24.66}$

2-180. $\dfrac{33.97 \times 10^{7}}{56.98 \times 10^{3}}$

2-181. $\dfrac{22{,}900. \times 10^{-6}}{76.4 \times 10^{4}}$

Combined multiplication and division

Since most scientific calculations involve both multiplication and division, the student should master the technique of combined multiplication and division. The projection rules for both multiplication and division also apply in a combination problem.

Example

(+2) (+4) (+6) - (+2) → +4

$$\frac{(513)(15{,}300)}{(238)} = 32{,}900, \text{ or } \mathbf{3.29 \times 10^4}$$

+2

ESTIMATION OF ANSWER BY SCIENTIFIC NOTATION:

$$\frac{5(10)^2(1.5)(10)^4}{(2.5)(10)^2} = \mathbf{3(10)^4} \leftarrow \text{ESTIMATED ANSWER}$$

In order to work the problem above, first set 513 divided by 238 on the C and D scales. Now, instead of reading this answer, move the hairline to 15,300 on the C scale (thus multiplying this latter quantity by the quotient of the first setting).

The student should always alternate the division and multiplication settings and should not try to take readings as he progresses with the steps. Only the final result is desired and since each reading of the rule further magnifies any error, the fewest readings possible should be allowed.

Example

(left extension from the division) ↓

(+1) (−4) (+2) (−1) - (+2) → 3 -1 = −4

$$\frac{(47.30)(0.000391)(693.5)}{(0.312)(55.1)(773.1)} = \mathbf{9.66 \times 10^{-4}}$$

(−1) (+1) (+2)

ESTIMATION OF ANSWER BY SCIENTIFIC NOTATION:

$$\frac{5(10^1\,4(10)^{-4}\,7(10)^2}{3(10)^{-1}\,6(10)^1\,8(10)^2} = \mathbf{1(10)^{-3}} \leftarrow \text{ESTIMATED ANSWER}$$

Remember that when you want to divide, you move the slide, and when you want to multiply, you move the hairline.

A common mistake made by many students is to multiply all the quantities in the dividend and all the quantities in the divisor and then divide these two results. This is a bad habit and such practice should not be followed. There are too many chances for mistakes, in addition to the method's being slower.

Combined multiplication and division practice problems

2-182. $\dfrac{(29.6)(18.01)}{(937)} = \mathbf{(5.69)(10)^{-1}}$

2-183. $\dfrac{(625{,}000)(0.0337)}{(48.2)} = \mathbf{(4.37)(10)^{2}}$

2-184. $\dfrac{(0.887)(1{,}109)}{(5.22)} = \mathbf{(1.884)(10)^{2}}$

2-185. $\dfrac{(0.1058)(937{,}000)}{(0.218)} = \mathbf{(4.55)(10)^{5}}$

2-186. $\dfrac{(43{,}800)(0.0661)}{(87.2 \times 10^{5})} = \mathbf{(3.32)(10)^{-4}}$

2-187. $\dfrac{(114.3)(0.567)}{(66{,}400)} = \mathbf{(9.76)(10)^{-4}}$

2-188. $\dfrac{(76.5 \times 10^{4})}{(0.733)(49.7 \times 10^{-6})} = \mathbf{(2.10)(10)^{10}}$

2-189. $\dfrac{(11.03)}{(20{,}100)(8.72 \times 10^{3})} = \mathbf{(6.29)(10)^{-8}}$

2-190. $\dfrac{(0.226)}{(87.3 \times 10^{4})(0.717)} = \mathbf{(3.61)(10)^{-7}}$

2-191. $\dfrac{(43.2)}{(9.09)(0.000652)} = \mathbf{(7.29)(10)^{3}}$

2-192. $\dfrac{(94.9 \times 10^{-9})}{(33{,}800)(0.609)} = \mathbf{(4.61)(10)^{-12}}$

2-193. $\dfrac{(737{,}000)}{(0.1556)(61.9 \times 10^{3})} = \mathbf{(7.65)(10)^{1}}$

2-194. $\dfrac{(17.01)(0.0336)}{(52{,}600)(0.01061)} \neq \mathbf{(1.024)(10)^{-3}}$

2-195. $\dfrac{(66.6)(0.937)}{(7.05 \times 10^{2})(184{,}300)} = \mathbf{(4.80)(10)^{-7}}$

2-196. $\dfrac{(2.96)(1000)(62.1)}{(0.911)(432{,}000)} = \mathbf{(4.67)(10)^{-1}}$

2-197. $\dfrac{(45.8)(10.33)}{(29{,}200)(0.702)} = \mathbf{(2.31)(10)^{-2}}$

2-198. $\dfrac{(0.604)(9{,}270)}{(0.817 \times 10^{4})(1.372)} = \mathbf{(4.99)(10)^{-1}}$

2-199. $\dfrac{(176{,}300)(42.8 \times 10^3)}{(68.3)(15.01)} = \mathbf{(7.36)(10)^6}$

2-200. $\dfrac{(39{,}200)(89.3 \times 10^{-7})}{(20.4 \times 10^{-6})(155.5)} = \mathbf{(1.104)(10)^2}$

2-201. $\dfrac{(0.763 \times 10^{-4})(0.01004)}{(44.3)(7{,}150{,}000)} = \mathbf{(2.42)(10)^{-15}}$

2-202. $\dfrac{(152{,}300)(88{,}100)}{(0.00339)(60.4)} = \mathbf{(6.55)(10)^{10}}$

2-203. $\dfrac{(90{,}400)(2.05 \times 10^6)}{(24.3 \times 10^{-2})(0.0227)} = \mathbf{(3.36)(10)^{13}}$

2-204. $\dfrac{(14.36 \times 10^2)(0.907)}{(51.6 \times 10^2)(0.00001118)} = \mathbf{(2.26)(10)^4}$

2-205. $\dfrac{(991{,}000)(60.3 \times 10^4)}{(23.3 \times 10^{-1})(0.1996)} = \mathbf{(1.285)(10)^{12}}$

2-206. $\dfrac{(8.40)(10)^3(29.6 \times 10^{-5})}{(0.369)(10.02 \times 10^9)} = \mathbf{(6.72)(10)^{-10}}$

2-207. $\dfrac{(54.9)(26.8)(0.331)}{(21.6)(11.03)(54.6)} = \mathbf{(3.74)(10)^{-2}}$

2-208. $\dfrac{(17{,}630)(0.1775)(92.3)}{(0.433)(0.0061)(57.3)} = \mathbf{(1.908)(10)^6}$

2-209. $\dfrac{(0.821)(0.221)(0.811)}{(0.0907)(10.72)(66{,}300)} = \mathbf{(2.28)(10)^{-6}}$

2-210. $\dfrac{(0.00552)(89.6)(0.705)}{(19.52 \times 10^3)(18.03)(22.4)} = \mathbf{(4.42)(10)^{-8}}$

2-211. $\dfrac{(30{,}600)(29.9)(0.00777)}{(485)(19.32)(62.6)} = \mathbf{(1.212)(10)^{-2}}$

2-212. $\dfrac{(54.1)(0.393)(16{,}070)}{(49.3 \times 10^3)(11.21)(61.6)} = \mathbf{(1.00)(10)^{-2}}$

2-213. $\dfrac{(44.2)(100.7)(62{,}400)}{(90.3)(75{,}100)(0.01066)} = \mathbf{(3.84)(10)^3}$

2-214. $\dfrac{(78.4)(15.59)(0.01669)}{(33.6)(88{,}100)(0.432)} = \mathbf{(1.595)(10)^{-5}}$

2-215. $\dfrac{(994{,}000)(21{,}300)(0.1761)}{(44.4)(71.2)(32.1 \times 10^4)} = \mathbf{3.67}$

2-216. $\dfrac{(16.21)(678{,}000)(56.6)}{(0.01073)(4{,}980)(30.3)} = \mathbf{(3.84)(10)^5}$

2-217. $\dfrac{(61.3 \times 10^3)(0.1718)(0.893)}{(21.6)(0.902)(0.01155)} = \mathbf{(4.18)(10)^4}$

2-218. $\dfrac{(20{,}900)(16.22 \times 10^4)(0.1061)}{(877)(20.1 \times 10^{-4})(5.03)} = \mathbf{(4.06)(10)^7}$

2-219. $\dfrac{(999{,}000)(17.33)(0.1562)}{(0.802)(0.0443)(29.3 \times 10^{-1})} = \mathbf{(2.60)(10)^7}$

2-220. $\dfrac{(16.21)(0.0339)(151.6)(0.211)}{(0.00361)(0.785)(93.2)(406)} = \mathbf{(1.640)(10)^{-1}}$

2-221. $\dfrac{(84.3)(0.916)(0.1133)(21.3)}{(66.2)(0.407)(55.3)(462)} = \mathbf{(2.71)(10)^{-4}}$

Problems

Solve by combined multiplication and division method:

2-222. $\dfrac{(0.916)}{(90.5)(13.06)}$

2-223. $\dfrac{(0.00908)}{(22.3)(33.2)}$

2-224. $\dfrac{(24.5)(43)}{(36)}$

2-225. $\dfrac{(82)(9.3)}{(56.5)}$

2-226. $\dfrac{(167)(842)}{(0.976)}$

2-227. $\dfrac{(5.72)(3690)}{(95.7)}$

2-228. $\dfrac{(925)(76.9)}{(37.6)}$

2-229. $\dfrac{(9.87)}{(1.76)(89)}$

2-230. $\dfrac{(85.4)}{(26.3)(213)}$

2-231. $\dfrac{(1525)}{(73.6)(0.007)}$

2-232. $\dfrac{(84{,}500)}{(126)(37.3)}$

2-233. $\dfrac{(76)(23.7)}{(13.5)(373)}$

2-234. $\dfrac{(6.23)(2.14)}{(0.00531)}$

2-235. $\dfrac{(21.3)(370)}{(10.9)(758)}$

2-236. $\dfrac{(0.00215)(2520)}{(7.57)(118)}$

2-237. $\dfrac{(755)(1.15)}{(51.4)(0.093)}$

2-238. $\dfrac{(916)(0.752)}{(5.16)}$

2-239. $\dfrac{(23.1)(1.506)}{(6.27)}$

2-240. $\dfrac{(42.6)(1.935)}{(750.3)}$

2-241. $\dfrac{(77.1)(10.53)}{(331.0)(73)}$

2-242. $\dfrac{(56.7)(0.00336)}{(15.06)(8.23)}$

2-243. $\dfrac{(14.5)(10)^3(6.22)}{(53.3)(0.00103)}$

2-244. $\dfrac{(42)(1000)}{(5.23)(0.00771)}$

2-245. $\dfrac{(1.331)}{(916)(506)}$

2-246. $\dfrac{(4320)(0.7854)}{(134)(0.9)}$

2-247. $\dfrac{(0.00713)(329)}{(0.0105)(1000)}$

2-248. $\dfrac{(103.4)(0.028)}{(0.0798)}$

2-249. $\dfrac{(1573)(4618)}{(3935)(97)}$

2-250. $\dfrac{(47.2)(0.0973)}{(85)(37.6)}$

2-251. $\dfrac{(0.0445)(0.0972)}{(0.218)(0.318)}$

2-252. $\dfrac{(39.1)(680{,}000)(3.52)(1.1 \times 10^6)}{(0.0316)(9.6 \times 10^6)(26.3)}$

2-253. $\dfrac{(7.69)(76{,}000)(5.63)(0.00314)}{(0.00365)(10 \times 10^6)}$

2-254. $\dfrac{(3.97)(6.71 \times 10^{-3})(0.067)}{(63.1)(3 \times 10^7)(7.61)(80{,}175)}$

2-255. $\dfrac{(697)(0.000713)(68.1)}{(234)(9.68)(5.1 \times 10^4)}$

2-256. $\dfrac{(43{,}400)(9.16)(8.1 \times 10^{-6})}{(0.00613)(67{,}000)(0.416)}$

2-257. $\dfrac{(691.6)(7.191)(3 \times 10^7)}{(410{,}000)(6.39)(0.0876)}$

2-258. $\dfrac{(37.615)(81.4)(9.687)(0.0017)}{(13.13)(0.076)(43)}$

2-259. $\dfrac{(51.2 \times 10^{-6})(3.41 \times 10^5)(36.1)}{(96.69)(7 \times 10^{-2})(0.134)}$

2-260. $\dfrac{(6.716)(3.2 \times 10^3)(0.0173)(413)}{(0.0000787)(6.6 \times 10^4)}$

2-261. $\dfrac{(1.061) \times 10^{-1})(96{,}000)(3.717)}{(7.34 \times 10^{-6})(3.9 \times 10^4)(13.5)}$

2-262. $\dfrac{(361)(482)(5.816)(38.91)(0.00616}{(0.07181)(3 \times 10^3)(39.36)}$

2-263. $\dfrac{(0.019) \times 10^8)(111.15)(0.0168)}{(7.96)(58.6)(0.0987)(3{,}000)}$

2-264. $\dfrac{(21.4)(0.82)(39.6 \times 10^{-1})}{(10.86)(6.7 \times 10^{-2})(37{,}613)}$

2-265. $\dfrac{(63{,}761)(43{,}890)(0.00761)}{(8 \times 10^6)(0.0781)(67.17)}$

2-266. $\dfrac{(516.7)(212 \times 10^3)(0.967)(34)}{(76{,}516)(2 \times 10^{-6})(618)}$

2-267. $\dfrac{(5.1 \times 10^8)(370)(8.71)(3{,}698)}{(0.00176)(36{,}170)}$

2-268. $\dfrac{(59.71 \times 10^{-6})(0.00916)(0.1695)(55.61)}{(17.33 \times 10^5)(0.3165)(10.56)(1.105)}$

2-269. $\dfrac{(773.6)(57.17)(0.316)(912.3)}{(56{,}000)(715{,}000)(471.3)}$

2-270. $\dfrac{(51.33)(461.3)(919)(5.03)}{(66{,}000)(71.52)(0.3316)(12.39)}$

2-271. $\dfrac{(0.6617)(75.391)(0.6577)(91.33)}{(0.3305)(5.69 \times 10)(0.00317 \times 10^{-5})}$

Proportions and ratios

A "ratio" of one number to another is the quotient of the first with respect to the second. For example, the ratio of a to b may be written as $a{:}b$ or $\frac{a}{b}$. A "proportion" is a statement that two ratios are equal. Thus, $2{:}3 = 6{:}B$ means that $\frac{2}{3} = \frac{6}{B}$.

The slide rule is quite useful in solving problems involving ratio or proportion because these fractions may be handled on any pair of matching identical scales of the rule. The C and D scales are most commonly used for this purpose.

In the example, $\frac{2}{3} = \frac{6}{B}$, 2, 3, and 6 are known values and B is unknown. The procedure to solve for B would be as follows:

1. Divide 2 by 3 (using the C and D scales). In this position the value 2 on the D scale would be located immediately beneath 3 on the C scale.

2. The equal ratio of $\frac{6}{B}$ would also be found on the C and D scales. The unknown value B may be read on the C scale immediately above the known value 6 on the D scale; $B = 9$.

With this particular location of the slide, every value read on the C scale bears the identical ratio of 2:3 to the number directly below it on the D scale. It is also important to remember that the cross products of a proportion are equal. In the above example, $3 \times 6 = 2 \times B$.

Examples

a. $\frac{47}{21} = \frac{18}{A}$ *Answer,* $A =$ **8.04**

b. $\frac{0.721}{1.336} = \frac{B}{89.3}$ *Answer,* $B =$ **48.2**

c. $\frac{15.9}{C} = \frac{72.1}{166.7}$ *Answer,* $C =$ **36.7**

d. $\frac{D}{0.1156} = \frac{0.921}{0.473}$ *Answer,* $D =$ **0.225**

e. $\frac{42{,}100}{7{,}060} = \frac{E}{0.0321}$ *Answer,* $E =$ **0.1914**

Folded scales

The CF and DF scales are called *folded scales.* They are identical with the C and D scales except that their indexes are in a different position. On the majority of slide rules, the CF and DF scales begin at the left end with the value π, which means that their indexes will be located near the center of the rule. On some rules the CF and DF scales may be folded at ϵ (2.718) or at some other number.

Since the CF and DF scales are identical in graduations with the C and D scales, they can be used in multiplication and division just as the C and D scales are. Another important fact may be noticed when the scales are examined; that is, if a number such as 2 on the C scale is set over a number such as 3 on the D scale, then 2 on the CF scale coincides with 3 on the DF scale. This means that operations may be begun or answers obtained on either the C and D scales or on the CF and DF scales.

For example, if we wish to multiply 2 by 6, and we set the left index of the C scale over 2 on the D scale, we observe that the product cannot be read on the D scale because 6 on the C scale projects past the right end of the rule. Ordinarily this would mean that the slide would need to be run to the left so that the right index of the C scale could be used. However, by using the folded scales, we notice that the 6 on the CF scale coincides with 12 on the DF scale, thereby eliminating an extra movement of the slide (see Figure 2–19). In many cases the use of the folded scales will reduce the number of times the slide must be shifted to the left because an answer would fall beyond the right end of the D scale.

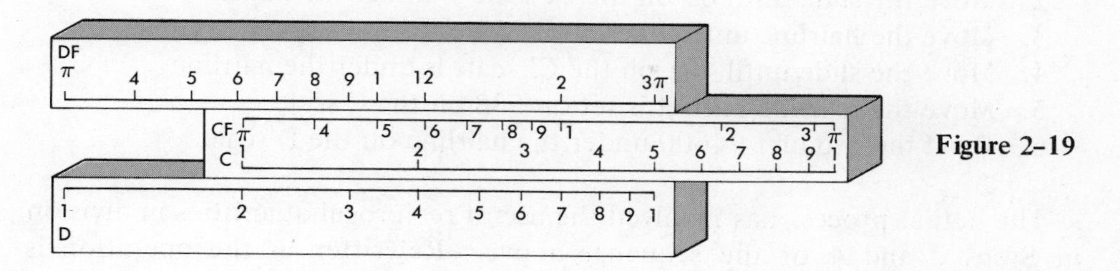

Figure 2–19

There are several methods by which the location of the decimal point in the answer can be determined. The decimal point location can best be found by using the method of scientific notation.

The projection rule can be used if it is always remembered that an answer read on the DF scale to the right of the index (near the center of the rule) corresponds to a left projection. Since in many operations the decimal point location in the answer can be determined by inspection, the decimal point can often be placed without reference to projection rules.

A convenient method of multiplying or dividing by π is afforded by the use of the folded scales. For example, to find the product 2π, set the hairline over 2 on the D scale. The product 6.28 is read on the DF scale under the hairline. Of course this same operation may be performed by using either index of the slide.

Reciprocal scales

The CI, DI, and CIF scales are known as *reciprocal scales* or *inverted scales.* They are identical with the C, D, and CF scales, respectively, except that they are inverted; that is, the numbers represented by the graduations on these scales increase from right to left. On some slide rules, the inverted scale graduations are printed in red to help distinguish them from the other scale markings.

An important principle to remember when using these scales is that a number on the C scale will have its reciprocal in the same position on the CI scale. Conversely, when the hairline is set to a number on the CI scale, its reciprocal is under the hairline on the C scale.

The inverted scales are useful in problems involving repeated multiplication or division because some movements of the slide may be eliminated.

Example Find the product:

$$(1.71)(8.30)(0.252)(4910)(53.8)$$

In order to perform this operation, using the inverted scales, the following steps are used:

1. Set the hairline to 1.71 on the D scale.
2. Move the slide until 83 on the CI scale is under the hairline.
3. Move the hairline until it is set on 252 on the C scale.
4. Move the slide until 491 on the CI scale is under the hairline.
5. Move the hairline until it is set on 538 on the C scale.
6. Read the product 94600 under the hairline on the D scale.

The actual process has involved the use of reciprocal quantities in division in Steps 2 and 4 of the sequence above. Rewritten as the operation is actually performed, the problem appears as follows:

$$\frac{(1.71)(0.252)(53.8)}{(1/8.30)(1/4910)}$$

ESTIMATION OF ANSWER BY SCIENTIFIC NOTATION:

$(2)(8)(2)(10)^{-1}(5)(10)^{3}(5)(10)^{1} = (8)(10)^{5}$ ← ESTIMATED ANSWER

Since the digits read on the slide rule were 945, the actual product would be $9.45(10)^5$. The projection rule should not be used with inverted scales, since the number of left projections are sometimes difficult to determine.

Proper use of the folded and inverted scales will enable one to work each practice problem below with only one setting of the slide.

Use of folded and reciprocal scales practice problems

2-272. $(264)(564)(522) = \mathbf{(7.77)(10)^7}$
2-273. $(387)(7.32)(176) = \mathbf{(4.99)(10)^5}$
2-274. $(0.461)(4.79)(1140) = \mathbf{(2.52)(10)^3}$
2-275. $(6.69)(1548)(92{,}000) = \mathbf{(9.53)(10)^8}$
2-276. $(561)(3.30)(1.94) = \mathbf{(3.59)(10)^3}$
2-277. $(1456)(0.351)(0.835) = \mathbf{(4.27)(10)^2}$
2-278. $(1262)(0.405)(65{,}100) = \mathbf{(3.33)(10)^7}$
2-279. $(0.1871)(5.04)(53{,}000) = \mathbf{(5.00)(10)^4}$
2-280. $(7.28 \times 10^{-5})(4.16)(14.10) = \mathbf{(4.27)(10)^{-3}}$
2-281. $(10.70)(19{,}400)(0.0914) = \mathbf{(1.897)(10)^4}$
2-282. $(4.56)(47.4)(87.1) = \mathbf{(1.883)(10)^4}$
2-283. $(0.510)(68.9)(3.370) = \mathbf{(1.184)(10)^5}$
2-284. $(2{,}030)(14.72)(129.7) = \mathbf{(3.88)(10)^6}$
2-285. $(1824)(29.1)(21{,}800) = \mathbf{(1.157)(10)^9}$
2-286. $(0.0255)(0.0932)(0.867) = \mathbf{(2.06)(10)^{-3}}$
2-287. $(93.6)(3.99)(5{,}680) = \mathbf{(2.12)(10)^6}$
2-288. $(4.48)(103.5)(0.198) = \mathbf{(9.18)(10)^1}$

2-289. $(0.580)(43{,}700)(40.3) = \mathbf{(1.021)(10)^6}$

2-290. $(7.05)(62.0)(34.9) = \mathbf{(1.526)(10)^4}$

2-291. $(74.8)(8.)(483{,}000) = \mathbf{(2.89)(10)^8}$

2-292. $\dfrac{(208)(90.2)}{(30{,}600)} = \mathbf{(6.13)(10)^{-1}}$

2-293. $\dfrac{(0.387)(25{,}200)}{(0.118)} = \mathbf{(8.26)(10)^4}$

2-294. $\dfrac{(0.458)(14.05 \times 10^{-15})}{(75.5 \times 10^8)} = \mathbf{(8.52)(10)^{-25}}$

2-295. $\dfrac{(18{,}100)(84.4)}{(10.92)} = \mathbf{(1.40)(10)^5}$

2-296. $\dfrac{(477)(9{,}720)}{(19{,}150)} = \mathbf{(2.42)(10)^2}$

2-297. $\dfrac{(25{,}600)}{(68{,}500)(12{,}080)} = \mathbf{(3.09)(10)^{-5}}$

2-298. $\dfrac{(3050)(1.00 \times 10^{-20})}{(71.4)(0.946)} = \mathbf{(4.52)(10)^{-19}}$

2-299. $\dfrac{(1{,}670)}{(0.000570)(24{,}700)} = \mathbf{(1.186)(10)^2}$

2-300. $\dfrac{(51.5)}{(15.14)(0.00194)} = \mathbf{(1.753)(10)^3}$

2-301. $\dfrac{(917{,}000)}{(54.3)(119.8 \times 10^{-4})} = \mathbf{(1.41)(10)^6}$

Squares and square roots

The A and B scales have been constructed so that their lengths are one half those of the C and D scales (see Figure 2–20). Similarly some slide rules are so constructed that they have a scale Sq 1 and Sq 2, or R_1 and R_2) which is twice as long as the D scale. This means that the logarithm of 3 as represented on the D scale would be equivalent in length to the logarithm of 9 on the A scale. Where the Sq 1 and Sq 2 or the R_1 and R_2 scales are used in conjunction with the D scale, the logarithm of 3 on the Sq 2 (R_2) scale would be equivalent in length to the logarithm of 9 on the D scale.

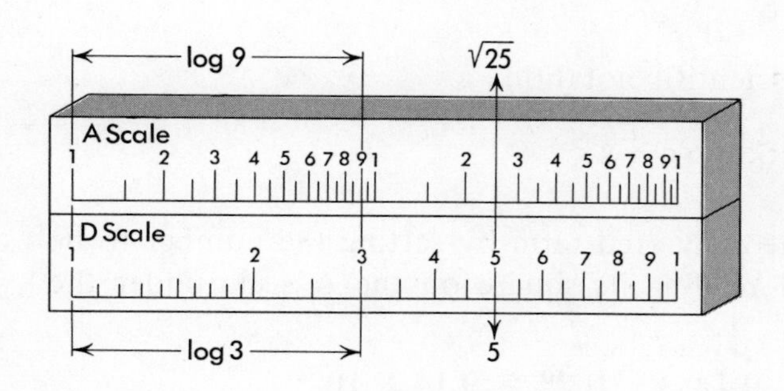

Figure 2–20

To find the square root of a number using the A and D scales

1. Get an estimate of the intended answer by placing a bar over every two digits, starting at the decimal point and working outward. There will be a digit in the answer for each bar marked.
2. Set the number on the A scale and read the square root on the D scale under the hairline. Note that the estimated answer will always indicate which A scale to use, since only one of the scales will give a square root near the estimated value.

Greater accuracy can be obtained by using the D scale in conjunction with the Sq 1 and Sq 2 scales (R_1 and R_2).

Examples for finding the location of decimal points:

a. $\overset{9\ \ x}{\sqrt{\overline{97}\ \overline{65}}}$ The estimated answer is somewhere between 90 and 100.

b. $\overset{.0\ \ 5}{\sqrt{.\overline{00}\ \overline{30}}}$ The estimated answer is approximately 0.05.

Note In the last example, since the given value was 0.003, an extra zero would have to be added after the 3 to complete the digits beneath the bar.

Examples for finding the square root of a number:

a. $\overset{1\ \ x\ \ x}{\sqrt{\overline{1}\ \overline{03}\ \overline{57}}}$ The estimated answer is somewhere between 100 and 200.

$\sqrt{\overline{1}\ \overline{03}\ \overline{57}} = 101.8 = \mathbf{1.018 \times 10^2}$

b. $\overset{0.0\ \ 2\ \ x}{\sqrt{0.\overline{00}\ \overline{05}\ \overline{20}}}$ The estimated answer is approximately 0.02.

$\sqrt{0.\overline{00}\ \overline{05}\ \overline{20}} = 0.02280 = \mathbf{2.280 \times 10^{-2}}$

Examples for finding squares:

1. Express the number in scientific notation.

   a. $(0.0000956)^2 = (9.56 \times 10^{-5})^2$

2. Square each part of the converted term by setting the number to be squared on the D scale and reading its square on the A scale under the hairline.

   a. $(9.56)^2 \times (10^{-5})^2 = 9.14 \times 10^{-10} = \mathbf{9.14 \times 10^{-9}}$

b. $(90100)^2 = (9.01 \times 10^4)^2$
$(9.01)^2 \times (10^4)^2 = 81 \times 10^8 = \mathbf{8.11 \times 10^9}$

c. $(357000000)^2 = (3.57 \times 10^8)^2$
$(3.57)^2 \times (10^8)^2 = 1.27 \times 10^{16} = \mathbf{1.27 \times 10^{17}}$

d. $(0.00000001050)^2 = (1.05 \times 10^{-8})^2$
$(1.05)^2 \times (10^{-8})^2 = \mathbf{1.10 \times 10^{-16}}$

Squares and square roots practice problems

2-302. $(408)^2 = \mathbf{(1{,}665)(10)^5}$
2-303. $(8.35)^2 = \mathbf{(6.97)(10)^1}$
2-304. $(3{,}980)^2 = \mathbf{(1.584)(10)^7}$
2-305. $(0.941)^2 = \mathbf{(8.85)(10)^{-1}}$
2-306. $(57.4)^2 = \mathbf{(3.29)(10)^3}$
2-307. $(0.207)^2 = \mathbf{(4.28)(10)^{-2}}$
2-308. $(784)^2 = \mathbf{(6.15)(10)^5}$
2-309. $(296{,}000)^2 = \mathbf{(8.76)(10)^{10}}$
2-310. $(1037)^2 = \mathbf{(1.075)(10)^6}$
2-311. $(8.93)^2 = \mathbf{(7.97)(10)^1}$
2-312. $(30.9)^2 = \mathbf{(9.55)(10)^2}$
2-313. $(43{,}300)^2 = \mathbf{(1.875)(10)^9}$
2-314. $(0.00609)^2 = \mathbf{(3.71)(10)^{-5}}$
2-315. $(0.846)^2 = \mathbf{(7.16)(10)^{-1}}$
2-316. $(55.2 \times 10^3)^2 = \mathbf{(3.05)(10)^9}$
2-317. $(0.0707)^2 = \mathbf{(5.00)(10)^{-3}}$
2-318. $(11.92 \times 10^{-4})^2 = \mathbf{(1.421)(10)^{-6}}$
2-319. $(0.291 \times 10^{-5})^2 = \mathbf{(8.47)(10)^{-12}}$
2-320. $(449{,}000)^2 = \mathbf{(2.02)(10)^{11}}$
2-321. $(0.000977)^2 = \mathbf{(9.55)(10)^{-7}}$
2-322. $(33.5 \times 10^{-6})^2 = \mathbf{(1.122)(10)^{-9}}$
2-323. $(8{,}810)^2 = \mathbf{(7.76)(10)^7}$
2-324. $(50.9 \times 10^6)^2 = \mathbf{(2.59)(10)^{15}}$
2-325. $(99{,}300)^2 = \mathbf{(9.86)(10)^9}$
2-326. $(0.0714 \times 10^{-6})^2 = \mathbf{(5.10)(10)^{-15}}$
2-327. $\sqrt{96{,}100} = \mathbf{(3.10)(10)^2}$
2-328. $\sqrt{0.912} = \mathbf{(9.55)(10)^{-1}}$
2-329. $\sqrt{24.9} = \mathbf{4.99}$
2-330. $\sqrt{0.01124} = \mathbf{(1.06)(10)^{-1}}$
2-331. $\sqrt{5{,}256} = \mathbf{(7.25)(10)^1}$
2-332. $\sqrt{0.3764} = \mathbf{(6.14)(10)^{-1}}$
2-333. $\sqrt{43{,}800{,}000} = \mathbf{(6.62)(10)^3}$
2-334. $\sqrt{0.01369} = \mathbf{(1.17)(10)^{-1}}$
2-335. $\sqrt{73.6} = \mathbf{8.58}$
2-336. $\sqrt{1.1025} = \mathbf{1.05}$
2-337. $\sqrt{487{,}000} = \mathbf{(6.98)(10)^2}$
2-338. $\sqrt{580.8} = \mathbf{(2.41)(10)^1}$
2-339. $\sqrt{0.00002767} = \mathbf{(5.26)(10)^{-3}}$
2-340. $\sqrt{0.1399} = \mathbf{(3.74)(10)^{-1}}$
2-341. $\sqrt{6{,}368} = \mathbf{(7.98)(10)^1}$
2-342. $\sqrt{1.142 \times 10^{-3}} = \mathbf{(3.38)(10)^{-1}}$
2-343. $\sqrt{6.496 \times 10^1} = \mathbf{8.06}$
2-344. $\sqrt{190{,}970} = \mathbf{(4.37)(10)^2}$
2-345. $\sqrt{3{,}204{,}000} = \mathbf{(1.79)(10)^3}$
2-346. $\sqrt{0.003807} = \mathbf{(6.17)(10)^{-2}}$
2-347. $\sqrt{0.08352} = \mathbf{(2.89)(10)^{-1}}$
2-348. $\sqrt{3069} = \mathbf{(5.54)(10)^1}$
2-349. $\sqrt{61.78 \times 10^{-4}} = \mathbf{(7.86)(10)^{-2}}$
2-350. $\sqrt{3.648 \times 10^{-8}} = \mathbf{(1.91)(10)^{-4}}$
2-351. $\sqrt{9.92 \times 10^5} = \mathbf{(9.96)(10)^2}$

Problems

Solve by method of squares and square roots.

2-352. $(1468.)^2$
2-353. $(0.886)^2$
2-354. $(67.4)^2$
2-355. $(11.96)^2$
2-356. $(0.00448)^2$
2-357. $(0.000551)^2$
2-358. $(9.22)^2$
2-359. $(64{,}800.)^2$
2-360. $(0.0668)^2$
2-361. $(16.85)^2$
2-362. $(1.802 \times 10^9)^2$
2-363. $(0.00358)^2$
2-364. $(5089)^2$
2-365. $(44{,}900.)^2$
2-366. $(64.88)^2$
2-367. $\sqrt{11.81}$
2-368. $\sqrt{4567.}$
2-369. $\sqrt{0.01844}$
2-370. $\sqrt{0.9953}$
2-371. $\sqrt{1395.}$

2-372. $\sqrt{0.0001288}$
2-373. $\sqrt{1.082 \times 10^2}$
2-374. $\sqrt{75.9}$
2-375. $\sqrt{\pi}$
2-376. $\sqrt{73{,}800.}$
2-377. $\sqrt{13.38}$
2-378. $\sqrt{93.07}$
2-379. $\sqrt{0.1148}$
2-380. $\sqrt{0.2776}$
2-381. $\sqrt{9.31}$

2-382. $(0.774)^2(11.47)^{1/2}$
2-383. $(0.1442)^{1/2}(33.89)^{1/2}$
2-384. $(54.23)^2(88{,}900)^{1/2}$
2-385. $\sqrt{234.5}\ \sqrt{55{,}900.}$
2-386. $\sqrt{16.38}\ \sqrt{45.6}\ \sqrt{0.9}$
2-387. $\sqrt{415.}\ \sqrt{\pi}\ \sqrt{86.4}$
2-388. $\sqrt{15.66}\ \sqrt{0.1904}\ \sqrt{\pi}$
2-389. $(34.77)^2(54.8)^2(0.772)^{1/2}$
2-390. $\sqrt{7.90}\ \sqrt{7.02}\ \sqrt{11.54}$
2-391. $\sqrt{31.19}\ \sqrt{56.7}\ \sqrt{54.8}$

Cubes and cube roots

The D and K scales are used to find the cube or cube root of a number as shown in Figure 2-21. The same general procedure is used as that followed for squaring numbers and taking the square root of a number. The K scale is divided into scales K_1, K_2, and K_3, which are each one third the length of the D scale. Thus, if a number is located on the D scale, the cube of the number will be indicated on the K scale. It follows that if a number is located on one of the K scales, the root of the number would appear on the D scale.

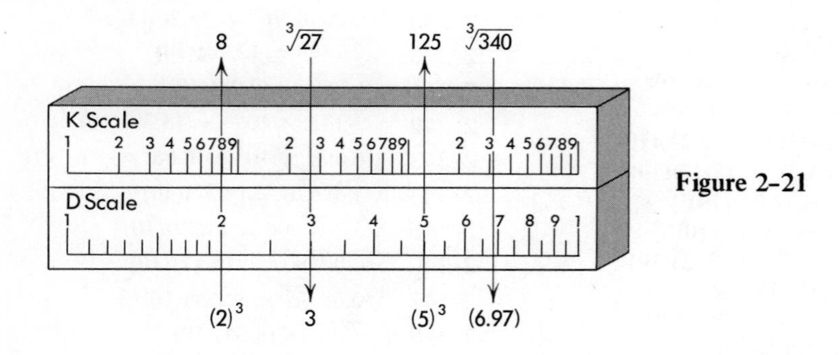

Figure 2-21

To find the cube root of a number

1. Get an estimate of the intended answer by placing a bar over every three digits, starting at the decimal point and working outward. There will be a digit in the answer for each bar marked.
2. Set the number on the K scale and read the cube root on the D scale under the hairline. (Some slide rules, such as those made by Pickett, have three cube root scales instead of the conventional K scale. These cube root scales are used with the D scale to determine cubes and cube roots of numbers. When they are used, however, the number should be set on the D scale and the cube root read on the appropriate cube root scale.)

Examples for finding the location of decimal points:

$$\begin{array}{l} \quad\ \ 3 \qquad x. \\ a.\ \sqrt[3]{\overline{44},\ \overline{800}.} \end{array}$$
The estimated answer is somewhere between 30 and 40.

$$\begin{array}{l} \quad\ \ 0. \quad\ 0 \quad\ 2 \\ b.\ \sqrt[3]{\overline{0}.\ \overline{000}\ \overline{011}} \end{array}$$
The estimated answer is approximately 0.02.

Note In estimating the answer by marking bars over the digit groupings, be sure that the bars cover three digits instead of two, as was the case in square roots.

Since an estimated answer [see Example *a* above] has been obtained, it is easy to pick the proper K scale (K_1, K_2, or K_3) to use. Remember that only one of these will give an answer between 30 and 40 [see Example *a*].

Examples for finding the cube roots of a number:

$$\begin{array}{l} \quad\ \ 1 \quad\ x \quad\ x \\ a.\ \sqrt[3]{\overline{1}\ \overline{490}\ \overline{000}.} \end{array}$$
The estimated answer is somewhere between 100 and 200.

$$\sqrt[3]{\overline{1}\ \overline{490}\ \overline{000}.} = 114.1 = \mathbf{(1.141)(10)^2}.$$

$$\begin{array}{l} \quad\ \ 0. \quad\ 0 \quad\ \ 6 \\ b.\ \sqrt[3]{\overline{0}.\ \overline{000}\ \overline{156}\ \overline{9}} \end{array}$$
The estimated answer is approximately 0.06.

$$\sqrt{\overline{0}.\ \overline{000}\ \overline{156}\ \overline{9}} = 0.0537 = \mathbf{(5.37)(10)^{-2}}.$$

Examples for finding cubes:

1. Convert the number to a number between 1 and 10 (scientific notation) that must be multiplied by 10 raised to some power.

   a. $(0.00641)^3 = (6.41 \times 10^{-3})^3$

2. Cube each part of the converted term by setting the number to be cubed on the D scale and reading its cube on the K scale under the hairline.

   a. $(6.41 \times 10^{-3})^3 = (264)(10)^{-9} = \mathbf{2.63 \times 10^{-7}}$

   b. $(93.88)^3 = (9.388 \times 10^1)^3$
   $(9.388)^3(10^1)^3 = 830 \times 10^3 = \mathbf{8.27 \times 10^5}$

   c. $(2{,}618{,}000.)^3 = (2.618 \times 10^6)^3$
   $(2.618)^3(10^6)^3 = (17.95 \times 10)^{18} = \mathbf{1.794 \times 10^{19}}$

   d. $(0.000001194)^3 = (1.194 \times 10^{-6})^3$
   $(1.194)^3(10^{-6})^3 = \mathbf{1.701 \times 10^{-18}}$

Cubes and cube roots practice problems

2-392. $(206)^3 = \mathbf{(8.74)(10)^6}$
2-393. $(7.68)^3 = \mathbf{(4.53)(10)^2}$
2-394. $(0.00519)^3 = \mathbf{(1.398)(10)^{-7}}$
2-395. $(33.5)^3 = \mathbf{(3.76)(10)^4}$
2-396. $(0.229)^3 = \mathbf{(1.201)(10)^{-2}}$
2-397. $(1090)^3 = \mathbf{(1.295)(10)^9}$
2-398. $(0.0579)^3 = \mathbf{(1.94)(10)^{-4}}$
2-399. $(9.89)^3 = \mathbf{(9.67)(10)^2}$
2-400. $(419)^3 = \mathbf{(7.36)(10)^7}$
2-401. $(52.4)^3 = \mathbf{(1.439)(10)^5}$
2-402. $(0.0249)^3 = \mathbf{(1.544)(10)^{-5}}$
2-403. $(14.9)^3 = \mathbf{(3.31)(10)^3}$
2-404. $(2.96)^3 = \mathbf{(2.59)(10)^1}$
2-405. $(397)^3 = \mathbf{(6.26)(10)^7}$
2-406. $(63.4)^3 = \mathbf{(2.55)(10)^5}$
2-407. $(9040)^3 = \mathbf{(7.39)(10)^{11}}$
2-408. $(0.0783)^3 = \mathbf{(4.80)(10)^{-4}}$
2-409. $(0.844)^3 = \mathbf{(6.01)(10)^{-1}}$
2-410. $(5.41)^3 = \mathbf{(1.583)(10)^2}$
2-411. $(35.5)^3 = \mathbf{(4.47)(10)^4}$
2-412. $(0.1270)^3 = \mathbf{(2.05)(10)^{-3}}$
2-413. $(20.7)^3 = \mathbf{(8.87)(10)^3}$
2-414. $(691)^3 = \mathbf{(3.30)(10)^8}$
2-415. $(0.719)^3 = \mathbf{(3.72)(10)^{-1}}$
2-416. $(4.34)^3 = \mathbf{(8.17)(10)^1}$
2-417. $\sqrt[3]{30{,}960{,}000} = \mathbf{(3.14)(10)^2}$
2-418. $\sqrt[3]{0.001728} = \mathbf{(1.20)(10)^{-1}}$
2-419. $\sqrt[3]{491} = \mathbf{7.89}$
2-420. $\sqrt[3]{9.91 \times 10^{11}} = \mathbf{(9.97)(10)^3}$
2-421. $\sqrt[3]{0.272} = \mathbf{(6.48)(10)^{-1}}$
2-422. $\sqrt[3]{118{,}400} = \mathbf{(4.91)(10)^1}$
2-423. $\sqrt[3]{22.91} = 2.84$
2-424. $\sqrt[3]{527{,}500} = \mathbf{(8.08)(10)^1}$
2-425. $\sqrt[3]{1.295} = \mathbf{1.09}$
2-426. $\sqrt[3]{0.0001804} = \mathbf{(5.65)(10)^{-2}}$
2-427. $\sqrt[3]{460{,}100{,}000} = \mathbf{(7.72)(10)^2}$
2-428. $\sqrt[3]{261{,}000} = \mathbf{(6.39)(10)^1}$
2-429. $\sqrt[3]{0.11620} = \mathbf{(4.88)(10)^{-1}}$
2-430. $\sqrt[3]{0.0030486} = \mathbf{(1.45)(10)^{-1}}$
2-431. $\sqrt[3]{0.03096} = \mathbf{(3.14)(10)^{-1}}$
2-432. $\sqrt[3]{504.4} = \mathbf{7.96}$
2-433. $\sqrt[3]{8{,}869{,}000} = \mathbf{(2.07)(10)^2}$
2-434. $\sqrt[3]{174{,}700{,}000} = \mathbf{(5.59)(10)^2}$
2-435. $\sqrt[3]{5.886 \times 10^{10}} = \mathbf{(3.89)(10)^3}$
2-436. $\sqrt[3]{5.885 \times 10^{-1}} = \mathbf{(8.38)(10)^{-1}}$
2-437. $\sqrt[3]{76.105 \times 10^{-5}} = \mathbf{(9.13)(10)^{-2}}$
2-438. $\sqrt[3]{327.1} = \mathbf{6.89}$
2-439. $\sqrt[3]{0.02567} = \mathbf{(2.95)(10)^{-1}}$
2-440. $\sqrt[3]{0.0004118} = \mathbf{(7.44)(10)^{-2}}$
2-441. $\sqrt[3]{68{,}420} = \mathbf{(4.09)(10)^1}$

Problems

Solve by method of cubes and cube roots.

2-442. $(86)^3$
2-443. $(148)^3$
2-444. $(395{,}000)^3$
2-445. $(47.6)^3$
2-446. $(1.074)^3$
2-447. $(76.9)^3$
2-448. $(220.8)^3$
2-449. $(9.72)^3$
2-450. $(110.7)^3$
2-451. $(91.3)^3$
2-452. $(1.757 \times 10^4)^3$
2-453. $(3.06 \times 10^{-7})^3$
2-454. $(44.8 \times 10^{-1})^3$
2-455. $(0.933 \times 10^{-2})^3$
2-456. $(0.1184 \times 10^8)^3$
2-457. $(51.5 \times 10^2)^3$
2-458. $\sqrt[3]{118}$
2-459. $\sqrt[3]{2{,}197}$
2-460. $\sqrt[3]{9}$
2-461. $\sqrt[3]{0.0689}$
2-462. $\sqrt[3]{0.001338}$
2-463. $\sqrt[3]{0.1794}$
2-464. $\sqrt[3]{0.0891}$
2-465. $\sqrt[3]{34{,}690.}$
2-466. $\sqrt[3]{0.3329}$
2-467. $\sqrt[3]{1{,}258{,}000}$
2-468. $\sqrt[3]{0.1853}$
2-469. $\sqrt[3]{12.88}$
2-470. $\sqrt[3]{4.98 \times 10^7}$
2-471. $\sqrt[3]{1.844 \times 10^{-5}}$
2-472. $\sqrt[3]{3.86 \times 10^{-1}}$
2-473. $(9.94)(0.886)^{1/3}$
2-474. $(284.)(11.98)^{1/3}$
2-475. $(0.117)(0.0964)^{1/3}$
2-476. $(\pi)^3(44.89)^3$
2-477. $(6.88)^3(0.00799)^3$

2-478. $(0.915)^{1/3}(0.366)^{1/3}\sqrt[3]{11{,}250}\,(36.12)^{1/3}$
2-479. $(2.34)^3(3.34)^3(4.56)^3(5.67)^3$
2-480. $(8.26)^{1/3}(8.26)^3(1000)^{1/3}(10)^3$
2-481. $\sqrt[3]{2670}\,\sqrt[3]{3165}\,\sqrt[3]{1065}\,\sqrt[3]{7776}$
2-482. $\sqrt[3]{206}\,\sqrt[3]{0.791}\,(12.35)^3(26.3)^3$

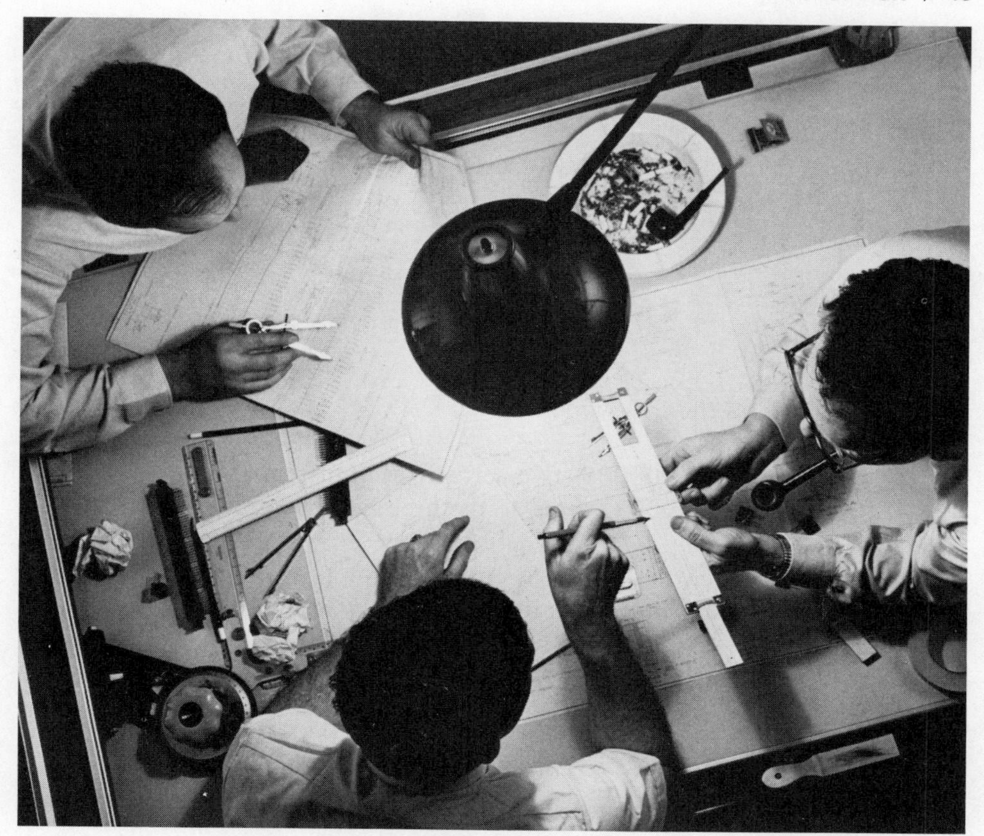

Illustration 2-3. The slide rule is particularly useful in the analysis of engineering designs. (Courtesy Sylvania Metals and Chemical Parts Division.)

Trigonometric functions

Finding trigonometric functions on a log-log rule is a rather simple process. The angle may be read on the S (sine), ST (sine and tangent of small angles), or T (tangent) scales. The functions may be read under the hairline on the C, D, or DI scales without any movement of the slide.

Sine 0° to 0.574° It is not often that the student needs to know the function of extremely small angles, but if he does need them, it is possible to get approximate values for these functions without consulting tables.

Method 1 (Based upon the relation that the sine of small angles is approximately equal to the size of the angle expressed in radians)

1. This method is more accurate than the following Method 2, and is preferable.
2. Express the angle in question in degrees.
3. Change the degrees to radians by dividing by 57.3.

Note 57.3° = 1 radian (approximately)

4. The value obtained is the approximate answer.

Example

$$\sin 6' = ?$$
$$6' = 6/60 = 0.10°$$
$$\sin 6' = \frac{0.10}{57.3}$$
$$\sin 6' = \mathbf{0.00174} \text{ approximately}$$

Method 2

1. Keep in mind the following values:

$\sin 1'' =$ **0.000005** (five zeros-five) approximately
$\sin 1' =$ **0.0003** (three zeros-three) approximately

2. For small angles, multiply the value of 1′ or 1″, as the case may be, by the number of minutes or seconds in question.

Example

$$\sin 6' = ?$$
$$\sin 6' = (6)(\sin 1')$$
$$\sin 6' = (6)(0.0003)$$
$$\sin 6' = \mathbf{0.0018} \text{ approximately}$$

Sine 0.574° to 5.74° To find the sine of an angle between 0.574° and 5.74°, the ST and D scales are used as shown in Figure 2–22.

Example $\sin 1.5° = ?$

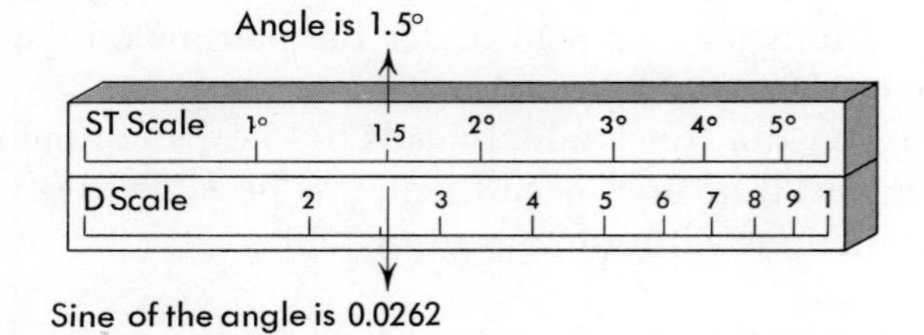

Figure 2–22

Instructions

1. Be certain that the left index of the D scale is directly under the left index of the ST scale.
2. Set the hairline to the angle on the ST scale.
3. Read the answer on the D scale. The answer will be a decimal number and will have one zero preceding the digits read from the rule.

Sine 5.74° to 90° To find the sine of an angle between 5.74° and 90°, the S and D scales are used as shown in Figure 2–23.

Example sin 45° = ?

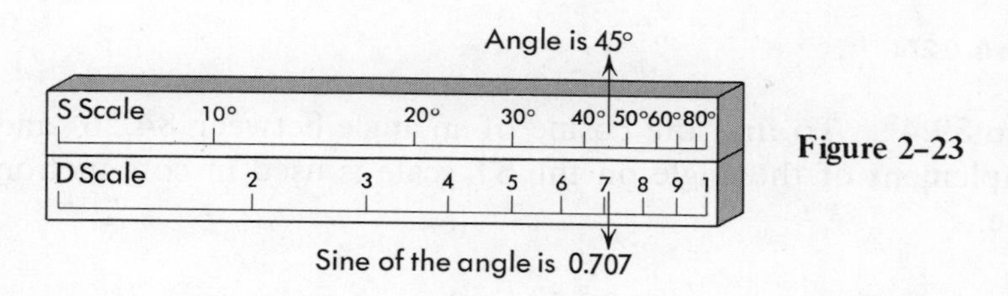

Figure 2–23

Instructions

1. Be certain that the left index of the D scale is directly under the left index of the S scale.
2. Set the hairline to the angle on the S scale. If the rule has more than one set of figures on the S scale, the angles for sine functions are usually shown to the right of the longer graduations.
3. Read the answer on the D scale. Place the decimal preceding the first digit read from the rule.

Sines practice problems

2-483. sin 26° = **0.438**
2-484. sin 81° = **0.988**
2-485. sin 16° = **0.276**
2-486. sin 15.5° = **0.267**
2-487. sin 42.6° = **0.677**
2-488. sin 3.33° = **0.0581**
2-489. sin 10.17° = **0.1765**
2-490. sin 63.2° = **0.893**
2-491. sin 70.83° = **0.945**
2-492. sin 26.67° = **0.449**
2-493. sin 7.33° = **0.1276**
2-494. sin 2.83 = **0.0494**
2-495. sin 51.5° = **0.783**
2-496. sin 5.17° = **0.0901**
2-497. sin 33.8° = **0.556**
2-498. sin 20.3° = **0.348**
2-499. sin 68.2° = **0.928**
2-500. arc sin 0.557 = **33.8°**
2-501. $\sin^{-1}$ 0.032 = **1.83°**
2-502. $\sin^{-1}$ 0.242 = **14.0°**
2-503. arc sin 0.709 = **45.15°**
2-504. $\sin^{-1}$ 0.581 = **35.5°**
2-505. arc sin 0.999 = **87.5°**
2-506. $\sin^{-1}$ 0.569 = **34.68°**
2-507. $\sin^{-1}$ 0.401 = **23.6°**

Cosine 0° to 84.26° To find the cosine of an angle between 0° and 84.26°, the markings to the left of the long graduations on the S scale are used in conjunction with the D scale. Note that the markings begin with 0° at the right end of the scale and progress to 84.26° at the left end of the scale as shown in Figure 2–24.

Example cos 74.1° = ?

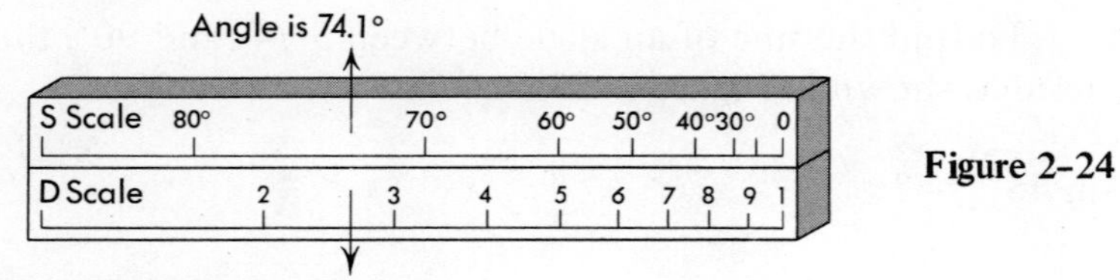

Figure 2-24

Cosine 84.26° to 89.4° To find the cosine of an angle between 84.26° and 89.4°, the complement of the angle on the ST scale is used in conjunction with the D scale.

Example

$$\begin{aligned} \cos 88.5° &= ? \\ \text{complement of } 88.5° &= 1.5° \\ \sin 1.5° &= 0.0262 \\ \cos 88.5° &= \mathbf{0.0262} \end{aligned}$$

Cosine 89.4° to 90° To find the cosine of an angle between 89.4° and 90°, determine the complement of the angle and find the value of the sine of this angle as previously discussed.

Example

$$\begin{aligned} \cos 89.94° &= ? \\ \text{complement of } 89.94° &= 0.06° \\ \sin 0.06° = \frac{0.06}{57.3} &= 0.001048 \\ \cos 89.94° &= \mathbf{0.001048} \end{aligned}$$

Note In finding the cosine of any angle, it is sometimes more convenient to look up the sine of the complement of the angle.

Example

$$\begin{aligned} \cos 60° &= ? \\ \text{complement of } 60° &= 30° \\ \sin 30° &= 0.500 \end{aligned}$$

Therefore,

$$\cos 60° = \mathbf{0.500}$$

Cosines practice problems

2-508. cos 18.8° = **0.947**
2-509. cos 33.17° = **0.837**
2-510. cos 71.5° = **0.317**
2-511. cos 45° = **0.707**
2-512. cos 68.3° = **0.370**
2-513. cos 26.9° = **0.892**
2-514. cos 55.7° = **0.564**
2-515. cos 5.5° = **0.995**
2-516. cos 81.3° = **0.151**
2-517. cos 8.9° = **0.988**
2-518. cos 77.6° = **0.215**
2-519. cos 39.1° = **0.776**
2-520. cos 50.7° = **0.633**
2-521. cos 11.5° = **0.980**
2-522. cos 49.2° = **0.653**
2-523. arc cos 0.901 = **25.7°**

2-524. $\cos^{-1} 0.727 =$ **43.4°**
2-525. $\cos^{-1} 0.0814 =$ **85.3°**
2-526. arc cos 0.284 = **73.5°**
2-527. $\cos^{-1} 0.585 =$ **54.2°**
2-528. $\cos^{-1} 0.658 =$ **48.8°**
2-529. $\cos^{-1} 0.1190 =$ **83.1°**
2-530. arc cos 0.303 = **72.4°**
2-531. $\cos^{-1} 0.505 =$ **59.7°**
2-532. $\cos^{-1} 0.693 =$ **46.1°**

Tangent 0° to 5.74° For small angles (0° to 5.74°) the tangent of the angle may be considered to be the same value as the sine of that angle.

Tangent 5.74° to 45° To find the tangent of an angle between 5.74° and 45°, the T scale is used in conjunction with the D scale, as shown in Figure 2-25.

Example Find tan 30°

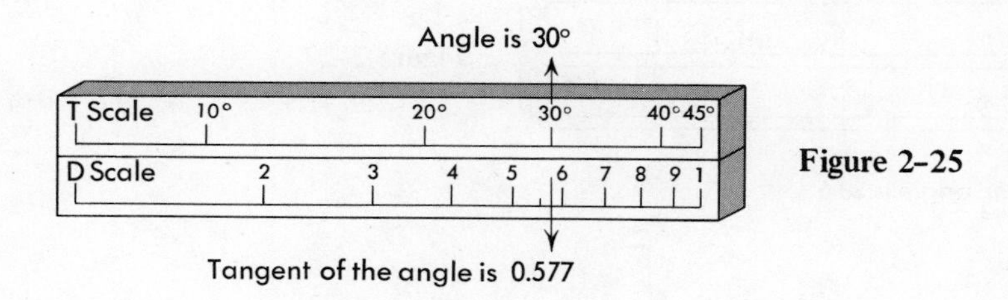

Figure 2-25

Instructions

1. Be certain that the left index of the D scale is directly under the left index of the T scale.
2. Set the hairline to the angle on the T scale. If the T scale has more than one set of markings, be certain that the correct markings are used.
3. Read the answer on the D scale. Place the decimal preceding the first digit read from the rule.

Tangent 45° to 84.26° To find the tangent of an angle between 45° and 84.26°, the markings to the left of the longer graduations on the T scale are used in conjunction with the CI or DI scales, as shown in Figure 2-26.

Example tan 70° = ?

Angle is 70°

T Scale 80° 70° 60° 50° 45°

9 8 7 6 5 4 3 2 DI Scale

Tangent of the angle is 2.74

Figure 2-26

Instructions

1. Be certain that the left index of the DI or CI scale is aligned with the left index of the T scale.
2. Set the hairline to the angle on the T scale.
3. Read the answer on the CI or DI scale. Note that these scales read from right to left. Place the decimal after the first digit read from the rule.

Tangent 84.26° to 89.426° To find the tangent of an angle between 84.26° and 89.426°, the complement of the angle on the ST scale is used in conjunction with the CI or DI scales, as shown in Figure 2-27.

Example tan 88° = ?

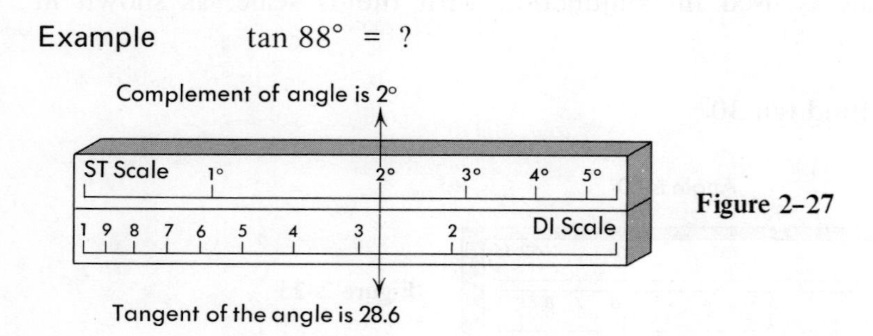

Figure 2–27

Instructions

1. Be certain that the left index of the DI or CI scale is aligned with the left index of the ST scale.
2. Complement of 88° = 2°.
3. Read the answer on the DI or CI scale. Note that these scales read from right to left.
4. Place the decimal point after the first two digits read from the rule.

Frequently the value of the function of an angle is known and it is desired to find the value of the angle.

Example $\sin \theta = 0.53;$
$\theta = ?$

This may be written in the inverse form in either of two ways:

$$\text{arc sin } 0.53 = \theta$$

or

$$\sin^{-1} 0.53 = \theta$$

then

$$\theta = 32°$$

The forms arc sin, arc cos, and arc tan are usually preferred in modern practice.

Tangent practice problems

2-533. tan 29.6° = **0.568**
2-534. tan 48.2° = **1.118**
2-535. tan 11.5° = **0.203**
2-536. tan 71.9° = **3.06**
2-537. tan 5.7° = **0.0993**
2-538. tan 61.4° = **1.834**
2-539. tan 33.3° = **0.657**
2-540. tan 69.2° = **2.63**
2-541. tan 40.6° = **0.857**
2-542. tan 8.7° = **0.1530**
2-543. tan 17.5° = **0.315**
2-544. tan 85.1° = **11.66**
2-545. tan 58.6° = **1.638**
2-546. tan 39.3° = **0.818**
2-547. tan 20.9° = **0.382**
2-548. tan 42.1° = **0.904**
2-549. arc tan 0.362 = **19.9°**
2-550. arc tan 0.841 = **40.1°**
2-551. $\tan^{-1}$ 0.119 = **6.78°**
2-552. $\tan^{-1}$ 0.0721 = **4.13°**
2-553. $\tan^{-1}$ 1.732 = **60°**
2-554. arc tan 21.6 = **87.3°**
2-555. $\tan^{-1}$ 0.776 = **37.8°**
2-556. arc tan 89.3 = **89.36°**
2-557. $\tan^{-1}$ 0.661 = **33.5°**

The following tables have been prepared for reference purposes. The student should check all the examples with his rule as he proceeds.

Function	*Angle*	*Read angle on*	*Read function on*	*Decimal*	*Examples*
sine or tangent	0°–0.574°	Convert the angle to radians (1 radian = 57.3°), and this value is assumed to be equal to the sine or tangent of the angle.			
sine or tangent	0.574°–5.74°	ST	D	0.0xxx	tan 2° = 0.0349 sin 3° = 0.0523
sine	5.74°–90°	S (right markings)	D	0.xxxx	sin 29° = 0.485
cosine	0°–84.26°	S (left markings)	D	0.xxxx	cos 43° = 0.7314
tangent	5.74°–45°	T (right markings)	D	0.xxxx	tan 13° = 0.231
tangent	45°–84.26°	T (left markings)	DI	x.xxx	tan 78° = 4.70
tangent	84.26°–89.426	Set complement on ST	DI	xx.xxx	tan 89° = 57.3
cosecant	5.74°–90°	S (right markings)	DI	x.xxx	csc 63° = 1.122
secant	0°–84.26°	S (left markings)	DI	x.xxx	sec 48° = 1.494
cotangent	0.574°–5.74°	ST	DI	xx.xx	cot 3.5° = 16.35
cotangent	5.74°–45°	T (right markings)	DI	x.xxx	cot 23° = 2.36
cotangent	45°–84.26°	T (left markings)	D	0.xxxx	cot 68° = 0.404

Trigonometric functions: problems

Solve, using the slide rule.

2-558. sin 35°
2-559. sin 14°
2-560. sin 78°
2-561. sin 3.7°
2-562. sin 88.3°
2-563. sin 55.3°
2-564. cos 35°
2-565. cos 66°
2-566. cos 21.3°
2-567. cos 11.1°
2-568. cos 7.9°
2-569. cos 43.8°
2-570. tan 33.8°
2-571. tan 9.4°
2-572. tan 37.7°
2-573. tan 22.5°
2-574. tan 86.1°
2-575. tan 54.4°
2-576. tan 70.3°
2-577. tan 29.7°
2-578. tan 36.5°
2-579. tan 13.3°
2-580. tan 45.8°
2-581. cot 14.7°
2-582. cot 81.8°
2-583. cot 36.9°
2-584. cot 61.2°
2-585. cot 54.3°
2-586. cot 18.7°
2-587. cot 3.77°
2-588. cot 66.4°
2-589. csc 38.1°
2-590. csc 75.2°
2-591. csc 88.3°
2-592. csc 12.8°
2-593. csc 46.4°
2-594. csc 81.1°
2-595. csc 32.6°
2-596. csc 9.03°
2-597. sec 6.14°
2-598. sec 59.2°
2-599. sec 79.4°
2-600. sec 19.5°
2-601. sec 2.77°
2-602. sec 45.9°
2-603. arc sin 0.771
2-604. arc cos 0.119
2-605. arc tan 34.8
2-606. arc sec 7.18
2-607. arc csc 1.05
2-608. cos 33.4°
2-609. cos 3.6°
2-610. arc cos 0.992
2-611. cos 24.67°
2-612. $\cos^{-1} 0.496$
2-613. cos 36°6′
2-614. arc cos 0.238
2-615. cos 0.75°
2-616. cos 36.6°
2-617. tan 32.6°
2-618. tan 16.34°
2-619. tan 88°30′
2-620. arc tan 0.62
2-621. $\tan^{-1} 0.75$
2-622. arc tan 0.392
2-623. $\tan^{-1} 1.53$
2-624. tan 37°24′
2-625. arc tan 0.567
2-626. $\tan^{-1} 0.0321$
2-627. cot 19°33′
2-628. sec 46°46′
2-629. csc 32°12′
2-630. sin 37°
2-631. sin 51°50′
2-632. sin 68°37′
2-633. sin 75°10′
2-634. arc sin 0.622
2-635. sin 13.6°
2-636. $\sin^{-1} 0.068$
2-637. sin 14.6°
2-638. arc sin 0.169
2-639. sin 34.67°
2-640. cos 26.26°
2-641. csc 20°20′
2-642. (csc 20°)(sin 46°)
2-643. (cos 32°)(tan 43°)

2-644. $\dfrac{(\sin 13.9°)}{(\cot 13.9°)}$

2-645. $\dfrac{\cot 33°22'}{\sec 4°53'}$

2-646. $\dfrac{(\cos 33°15')}{(\cot 46°19')}$

2-647. $\dfrac{(\sec 10°)(\cot 10°)}{(\sin 10°)(\csc 10°)}$

2-648. $\dfrac{(\sin 35°)(\tan 22°)}{(\sqrt[3]{\sin 5.96°})}$

2-649. $\dfrac{(\sec 11°)(\tan 4°)}{(\cot 49°)}$

2-650. $\dfrac{(\sin 8°)(\tan 9°)}{(\cot 82°)}$

2-651. $\dfrac{(\sin 1.36°)(\cot 26°)}{(\sqrt[3]{0.00916})}$

2-652. $\dfrac{\cot \sin^{-1} 0.916}{(1.32)(5.061)}$

2-653. $\dfrac{(77.19)(\sec 46°)}{(\tan 3.91°)}$

2-654. $\dfrac{(\sqrt[3]{\tan 25.9°})(\sin \cos^{-1} 0.5)}{(\sin 5.16°)(\tan 22°)}$

2-655. $\dfrac{(0.0311)(\sec 69°)\sqrt[3]{9.0}}{(\sin 9°)(\cos 9°)}$

2-656. $\dfrac{(1.916)(\sqrt[3]{1.916})(\sqrt[3]{\sin 20°})}{(\sqrt[3]{\sec 40°})(\tan 10°22')}$

2-567. $\dfrac{(6.17)(\tan 6.17°)(\sqrt[3]{6.17})}{(6.17)^2(\sin 61.7°)(\cos 6.17°)}$

Right triangle solution (log-log rule)

In the study of truss design, moments, and free body diagrams, the right triangle plays an important role. Since the Pythagorean theorem is sometimes awkward to use, and mistakes in arithmetic are likely to occur, it is suggested that the following method be used to solve right triangles.

Given: Right triangle with sides a, b, and c and angles A, B, and C (90°), as shown in Figure 2–28.

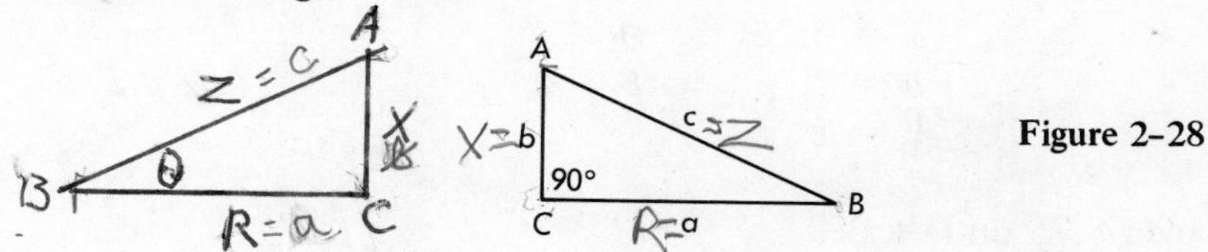

Figure 2–28

If the smaller side (b) is divided by the longer side (a) and the quotient is greater than 0.100, use *Solution 1.* If the quotient is between 0.100 and 0.0100, use *Solution 2.* If the quotient is less than 0.0100, assume that the hypotenuse (c) is equal in length to the longest side (a) and that $B \cong 0°$.

Solution 1

1. Set the index of the T scale above the larger side (a) on the D scale.
2. Move the hairline to the smaller side (b) on the D scale.
3. Read the two angles of the right triangle on the T scale. The larger angle is always opposite the larger side.
4. Move the slide until the smaller of the two angles just read is under the hairline on the sine scale.
5. Read the hypotenuse (c) on the D scale as indicated by the index of the sine scale.

Example

$a = 4$ $\qquad A = ?$
$b = 3$ $\qquad B = ?$
$c = ?$

a. Set right index of T to 4 on the D scale.
b. Move the hairline to 3 on the D scale.
c. Read $B = 36.9°$, $A = 53.1°$ on the T scale. (Note that the smaller angle is opposite the smaller side.)
d. Move the slide so that 36.9° on the S scale is under the hairline.
e. Read side $c = 5$ at the right index of the S scale on the D scale.

Solution 2

1. Set the index of the T scale above the largest side (a) on the D scale.
2. Move the hairline to the smaller side (b) on the D scale.

3. Read the smaller angle (B) on the ST scale. The other angle (A) is the complement of B.
4. The hypotenuse is assumed to be equal in length to the largest side.

Solution 3

This solution is used where the hypotenuse and one side are given.

Example

$a = 5.26$ $\quad A = ?$
$b = ?$ $\quad B = ?$
$c = 8.75$

a. Set index over 8.75 on D scale.
b. Move hairline to 5.26 on D scale.
c. Read $A = 37.0°$; $B = 53.0°$ on the S scale. (Note that the angle read on the sine scale is opposite the given side.)
d. Set hairline to 37° on the cosine scale.
e. Read $b = 7.0$ on the D scale.

Problems

Solve by right triangle method.

Problem	Given		Find		Problem	Given		Find	
2-658.	$a = 53$	$b = 4$	$B = ?$	$c = ?$	2-668.	$a = 11.33$	$B = 26.1°$	$b = ?$	$c = ?$
2-659.	$a = 69.3$	$c = 95$	$b = ?$	$A = ?$	2-669.	$a = 0.00197$	$A = 11.36°$	$b = ?$	$c = ?$
2-660.	$a = 37$	$c = 40.3$	$b = ?$	$B = ?$	2-670.	$c = 1904$	$A = 18.33°$	$a = ?$	$b = ?$
2-661.	$a = 1.97$	$c = 2.33$	$B = ?$	$b = ?$	2-671.	$c = 4.0059$	$B = 86.3°$	$a = ?$	$b = ?$
2-662.	$a = 29.3$	$c = 55.3$	$b = ?$	$A = ?$	2-672.	$c = 4.266$	$B = 31.06°$	$a = ?$	$b = ?$
2-663.	$a = 49.3$	$b = 29.6$	$c = ?$	$A = ?$	2-673.	$a = 0.00397$	$c = 0.00512$	$b = ?$	$A = ?$
2-664.	$a = 57.3$	$b = 42.1$	$c = ?$	$A = ?$	2-674.	$a = 1069$	$A = 85.3°$	$b = ?$	$c = ?$
2-665.	$a = 3.95$	$b = 1.06$	$c = ?$	$B = ?$	2-675.	$b = 42.1$	$B = 3.56°$	$a = ?$	$c = ?$
2-666.	$a = 333$	$b = 20$	$A = ?$	$c = ?$	2-676.	$a = 0.0317$	$c = 0.0444$	$b = ?$	$B = ?$
2-667.	$a = 591$	$b = 25$	$c = ?$	$B = ?$	2-677.	$a = 21.67$	$b = 20.06$	$c = ?$	$B = ?$

The log-log (Lon) scales

There are two groups of log-log scales (also called "Lon" scales) on the slide rule. Scales within the two groups are arranged in matched sets. Some slide rules have four matched sets, whereas others have three. These scales are used to obtain the roots, powers, and logarithms of numbers. The matched sets are arranged as follows:

Matched sets of log-log scales

Four sets		*Three sets*	
For numbers larger than one (called "Lon" scales)	*For numbers smaller than one (called "Lon-minus" scales)*	*(called LL scales)*	*(called LL_0 scales)*
Ln0	Ln-0		
Ln1	Ln-1	LL_1	LL_{01}
Ln2	Ln-2	LL_2	LL_{02}
Ln3	Ln-3	LL_3	LL_{03}

The C and D scales are used in conjunction with these matched sets of log-log scales. In former years other rules were manufactured with only two LL_0 scales, and these are marked LL_0 and LL_{00}. The A and B scales were used with LL_0 and LL_{00} scales on this type of rule. The general principles discussed below apply to all of the varous types of log-log scales.

Scale construction

If the Lon scales Ln0, Ln1, Ln2, and Ln3 were placed end to end, they would form a continuous scale, as shown in Figure 2-29. Similarly, if the Lon-minus scales Ln-0, Ln-1, Ln-2, Ln-3 were placed end to end, they would form a continuous scale. The Lon-minus scales are graduated from approximately 0.999 to 0.00003 (representing the values of $\epsilon^{-0.001}$ to ϵ^{-10}). The Lon scales are graduated from approximately 1.001 to 22,026 (representing the values of $\epsilon^{0.001}$ to ϵ^{10}). Since $\epsilon^0 = 1$, values on both the Lon and Lon-minus scales approach the value 1.0000.

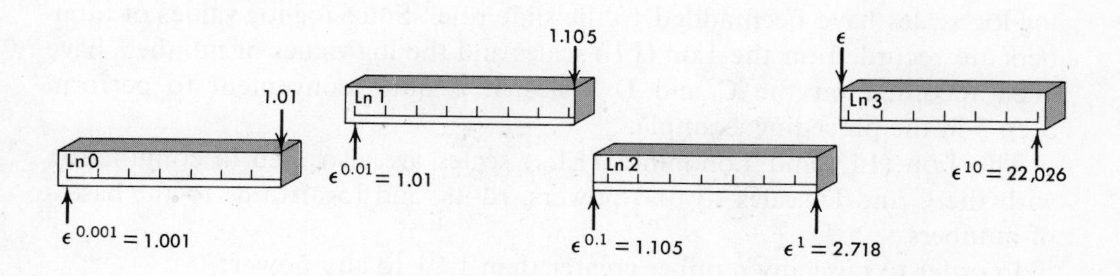

Figure 2-29

Each division on the Lon and Lon-minus scales represents a single unique number. Thus the decimal point is already marked on these scales for all of the numbers located on the scales. For example, there is only one place on the Lon scales that the number 125.0 may be found. The number 125.0 is found on the Ln3 (LL_3) scale, whereas the number 1.25 is found on the Ln2 (LL_2) scale. Since the manner in which settings are read on the log-log scales is distinctly different from the method of reading the scales previously studied, the student should be very careful in making his slide rule settings.

Reciprocal values

The only case where the Lon and Lon-minus (LL and LL_0) scales may be used together is in the finding of reciprocals of numbers. The reciprocal of any number on the Lon (LL) scales can be read on the corresponding Lon-minus (LL_0) scale.

Examples

1. Find 1.25 on Ln2 (LL_2) scale. On the Ln–2 (LL_{02}) scale its reciprocal can be read as 0.80.
2. Find 236 on the Ln3 (LL_3) scale. On the Ln–3 (LL_{03}) scale its reciprocal can be read as 0.00424.

Raising a number to a power

If such problems as $(5.3)^3 = ?$ were worked entirely by logarithms, the following procedure would be required:

1. $(5.3)^3 = ?$
2. $\log \text{ans.} = 3(\log 5.3)$
3. $\log [\log \text{ans}] = \log 3 + \log (\log 5.3)$
4. Answer = $(1.488)(10)^2$

Step 3 is rather involved in many instances. It is for this reason that the log-log scales have been added to the slide rule. Since log-log values of numbers are recorded on the Lon (LL) scales and the log values of numbers have been recorded on the C and D scales, it is quite convenient to perform Step 3 in the preceding example.

The Lon (LL) and Lon-minus (LL_0) scales are also used in conjunction with the C and D scales to find powers, roots, and logarithms to the base ϵ of numbers.

In order to raise any number greater than 1.01 to any power:

$$(X)^n = A$$

1. Set the index of the C scale over the value X found on the appropriate Lon (LL) scale (Ln0, Ln1, Ln2, or Ln3).
2. Move the hairline to the value n on the C scale.
3. Read the answer A on the appropriate Lon (LL) scale.

Example   As shown in Figure 2–30,

$$\begin{aligned} (1.02)^{2.5} &= ? \\ \log[\log \text{ans.}] &= \log 2.5 + \log(\log 1.02) \\ \text{Answer} &= \mathbf{1.0507} \end{aligned}$$

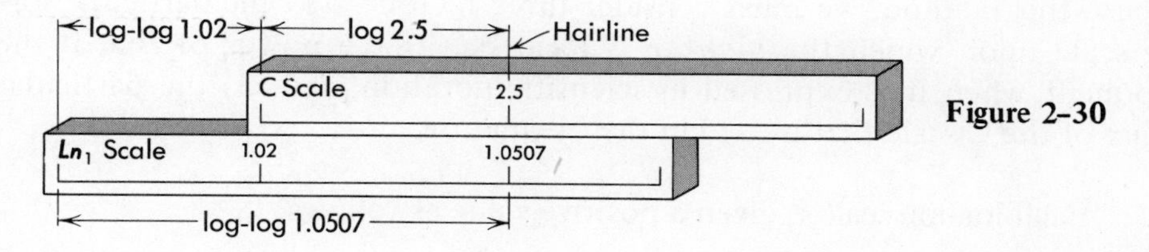

Figure 2–30

Solution

1. Set the index of the C scale over the value 1.02 on the Ln1 (LL_1) scale.
2. Move the hairline to the value 2.5 on the C scale.
3. Read the answer 1.0507 on the Ln1 (LL_1) scale.

These scales are arranged so that a number on the Ln3 (LL_3) scale is the tenth power of the number directly below it on the Ln2 (LL_2) scale, and the Ln2 (LL_2) scale gives the tenth power of a number in the corresponding position on the Ln1 (LL_1) scale. Therefore the Ln3 (LL_3) scale would give the one-hundredth power of a number in the corresponding position on the Ln1 (LL_1) scale.

Example $(1.034)^{0.23}$ = **1.00773** ans. on the Ln0
$(1.034)^{2.3}$ = **1.0799** ans. on the Ln1 (LL_1)
$(1.034)^{23.}$ = **2.156** ans. on the Ln2 (LL_2)
$(1.034)^{230.}$ = **2160** ans. on the Ln3 (LL_3)

In order to raise any number less than 0.99 to any power:

$$(X)^n = A$$

1. Set the index of the C scale over the value X found on the appropriate Lon-minus (LL_0) scale Ln-0, Ln-1, Ln-2, or Ln-3 (LL_{01}, LL_{02}, LL_{03}).
2. Move the hairline to the value n on the C scale.
3. Read the answer A on the appropriate Ln-0 (LL_0) scale.

Example $(0.855)^{4.8} = A$, as shown in Figure 2–31.

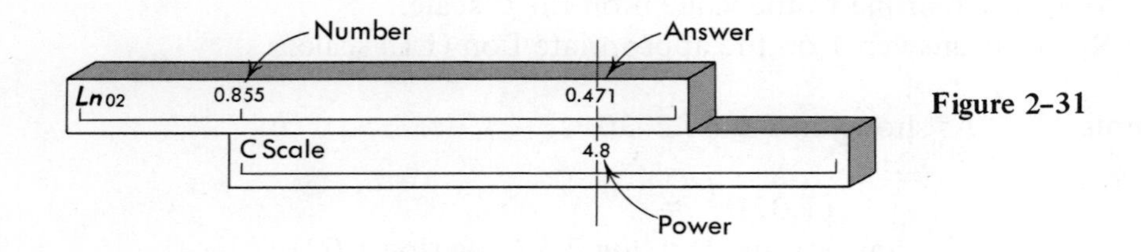

Figure 2–31

Method of scale selection—powers of numbers

To use this method, we must consider three factors: (1) the particular log-log scale upon which the *number* is located, (2) the power of ten of the exponent when it is expressed in scientific notation, and (3) the particular index of the C scale that is used in the calculation.

1. Each log-log scale is given a positive value as follows:

Ln0 = 0	Ln-0 = 0
Ln1 = + 1 (Also LL_1)	Ln-1 = + 1 (Also LL_{01})
Ln2 = + 2 (Also LL_2)	Ln-2 = + 2 (Also LL_{02})
Ln3 = + 3 (Also LL_3)	Ln-3 = + 3 (Also LL_{03})

2. The exponent should be expressed in scientific notation and the power of ten indicated.
3. Assume that the *left* index of the C scale has a value of zero (0) and that the *right* index has a value of plus one (+1).

Rule for scale selection of powers of numbers

The number of the scale upon which the answer will be read is the algebraic sum of (1) the value of the scale on which the number to be raised is found plus (2) the C scale index value plus (3) the power of ten of the exponent.

Example $(1.015)^{56} = ?$
Rewrite as $(1.015)^{56(10)1} = ?$

Factor	*Description of factor*	*Value*	
1.015	1.015 is found on LL_1 scale	+ 1	
Left Index	Use left index of C scale	0	
56	Power of ten of exponent = 1	+ 1	
?	Sum = Scale location of answer	+ 2	← Answer on Ln2 (LL_2)

Therefore, the answer will be read under the hairline on the Ln2 (LL_2) scale.

$(1.015)^{56} = 2.30$ Answer, as shown in Figure 2–32.

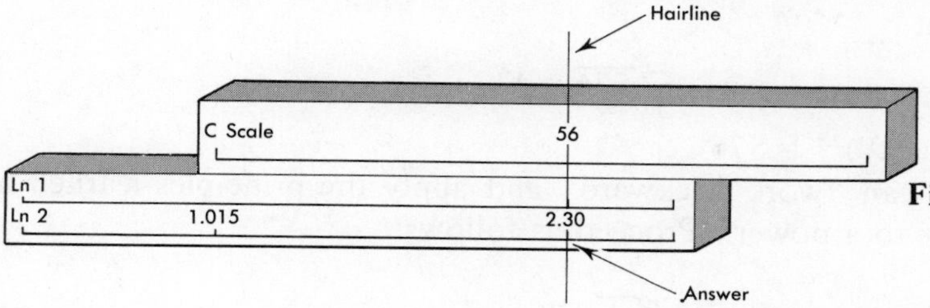

Hairline
C Scale
56
Ln 1
Ln 2
1.015
2.30
Answer

Figure 2–32

Negative exponents

In solving problems which involve raising numbers to a negative power, either of two methods may be employed.

Method 1 Set the number and its exponent on the proper scales in the usual manner. Instead of reading the answer on the usual log-log scale, read it on the corresponding scale of the other group.

Example $(9.2)^{-3.5} = ?$

Instead of reading the answer as 2355 on the Ln3 (LL_3) scale, read its reciprocal value on the Ln-3 (LL_{03}) as 0.000425; therefore

$$(9.2)^{-3.5} = \mathbf{4.25 \times 10^{-4}} \text{ (Answer)}$$

Method 2 Set the numbers on the rule in the usual manner, ignoring the negative exponent. When the answer by this operation has been obtained, determine its reciprocal, using the CI scale.

On the slide rules that have only the LL_0 and LL_{00} scales, Method 2 is the only method that can be used.

Powers of numbers: practice problems

2-678. $(53.2)^{0.84}$ = **28.2**
2-679. $(4.65)^{3.68}$ = **285.**
2-680. $(0.836)^{0.47}$ = **0.919**
2-681. $(1.0042)^{217}$ = **2.48**
2-682. $(0.427)^{4}$ = **0.0332**
2-683. $(0.3156)^{4}$ = **0.00992**
2-684. $(0.159)^{0.67}$ = **0.292**
2-685. $(1.0565)^{49.5}$ = **15.2**
2-686. $(32.5)^{0.065}$ = **1.254**
2-687. $(3.45)^{4.65}$ = **317.**
2-688. $(0.759)^{5}$ = **0.252**
2-689. $(2.127)^{4}$ = **20.5**
2-690. $(2.03)^{-5}$ = **0.0290**
2-691. $(4.00)^{0.0157}$ = **1.022**
2-692. $(0.0818)^{-0.777}$ = **7.00**
2-693. $(1.382)^{21.3}$ = **984.**
2-694. $(0.071)^{-0.46}$ = **3.38**
2-695. $(0.232)^{0.0904}$ = **0.876**
2-696. $(2.718)^{0.405}$ = **1.50**
2-697. $(0.916)^{0.724}$ = **0.9384**
2-698. $(1.1106)^{1.72}$ = **1.197**
2-699. $(59.2)^{-0.43}$ = **0.1729**
2-700. $(883)^{0.964}$ = **692.**
2-701. $(7676)^{0.001102}$ = **1.0099**
2-702. $(4.30)^{0.521}$ = **2.14**

Finding roots of numbers

The process of finding roots of numbers is easier to understand if it is remembered that

$$\sqrt[2.1]{576} = X$$

may be written as $(X)^{2.1} = 576$

Therefore we can "work backward" and apply the principles learned in raising a number to a power. Proceed as follows:

Example $$\sqrt[n]{A} = X$$

1. Locate the root n on the C scale to coincide with the value A found on the appropriate log-log scale.
2. Move the hairline to the particular index of the C scale which is located within the body of the rule.
3. Read the answer on the appropriate log-log scale.

Example $\sqrt[3.2]{120} = 4.46$ ans. on Ln3 (LL_3), as shown in Figure 2–33.

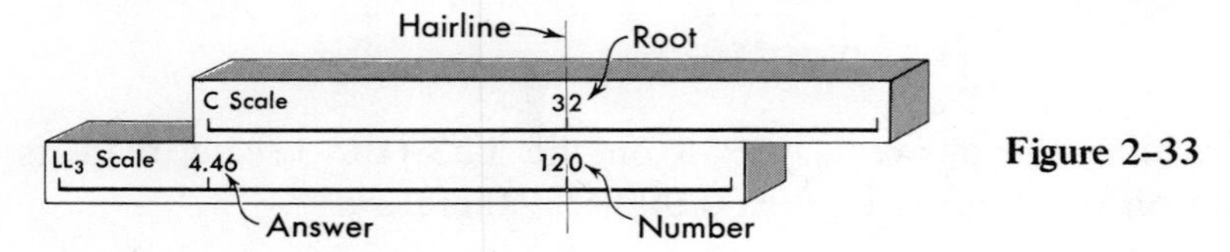

Figure 2–33

Also $\sqrt[32]{120} =$ **1.1615** ans. on Ln2 (LL_2)
$\sqrt[320]{120} =$ **1.0152** ans. on Ln1 (LL_1)

In taking the root of a number, students usually are less certain of the appropriate scale upon which the answer is found. Therefore, a method of scale selection similar to that employed for powers of numbers should be used.

Method of scale selection for roots of numbers

As before there are three factors which must be considered: (1) the particular log-log scale upon which the *number* is located; (2) the power of ten of the exponent when it is expressed in scientific notation, and (3) the particular index of the C scale which is used in the calculation.

1. Each log-log scale is given a negative value as follows:

Ln0 = 0	Ln-0 = 0
Ln1 = –1 (Also LL_1)	Ln-1 = –1 (Also LL_{01})
Ln2 = –2 (Also LL_2)	Ln-2 = –2 (Also LL_{02})
Ln3 = –3 (Also LL_3)	Ln-3 = –3 (Also LL_{03})

2. The root should be expressed in scientific notation and the power of ten indicated.

3. Assume that the *left* index of the C scale has a value of zero (0) and that the *right* index has a value of plus one (+1).

Rule for scale selection for roots of numbers

The number of the scale upon which the answer will be read is the algebraic sum of (1) the value of the scale on which the *number whose root is to be determined* is located, plus (2) the C scale index value, plus (3) the power of ten of the root.

Example $\sqrt[4.37]{0.0092}$ = Answer, as shown in Figure 2–34.

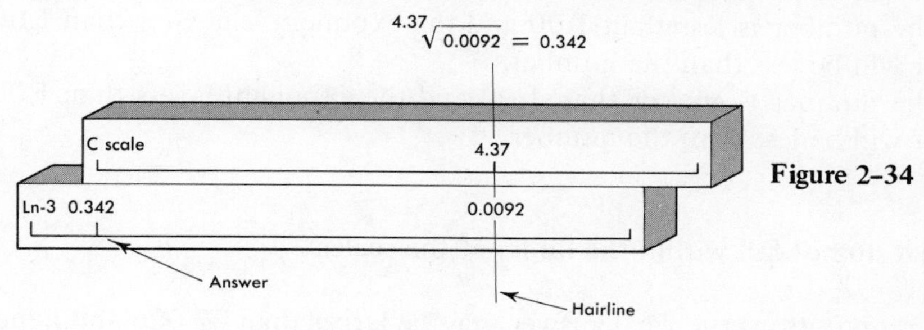

Figure 2–34

Factor	*Description of factor*	*Value*
0.0092	0.0092 is found on Ln-3 (LL_{03}) Scale	-3
Left Index	Use left index of C scale	0
4.37	Power of ten of root = 0	0
?	Sum = Scale location of Answer	-3 ← Answer on Ln-3 (LL_{03})

Therefore, the answer will be read on the Ln–3 (LL_{03}) scale as **0.342**.

Roots of numbers practice problems

2-703. $\sqrt[7.81]{5.85}$ = **1.254**
2-704. $\sqrt[6]{0.0835}$ = **0.661**
2-705. $\sqrt[5]{0.0763}$ = **0.598**
2-706. $\sqrt[194]{460.}$ = **1.0321**
2-707. $\sqrt[6]{0.0001}$ = **0.215**
2-708. $\sqrt[1.65]{8.26}$ = **3.60**
2-709. $\sqrt[0.34]{0.862}$ = **0.646**
2-710. $\sqrt[2.3]{85.9}$ = **6.93**
2-711. $\sqrt[60]{45.}$ = **1.0655**
2-712. $\sqrt[21.5]{1.606}$ = **1.0223**
2-713. $\sqrt[1.91]{92.5}$ = **10.7**
2-714. $\sqrt[50]{0.05}$ = **0.9418**
2-715. $\sqrt[7]{0.0108}$ = **0.524**
2-716. $\sqrt[0.006]{0.9762}$ = **0.018**
2-717. $\sqrt[5.21]{2000}$ = **4.30**
2-718. $\sqrt[0.04]{0.9792}$ = **0.591**
2-719. $\sqrt[2.7]{81}$ = **5.09**
2-720. $\sqrt[2.81]{1.218}$ = **1.0727**
2-721. $\sqrt[2.15]{52.5}$ = **6.31**
2-722. $\sqrt[400]{100}$ = **1.0116**
2-723. $\sqrt[0.75]{2.37}$ = **3.16**
2-724. $\sqrt[0.073]{1.060}$ = **2.22**
2-725. $\sqrt[1.51]{6.50}$ = **3.45**
2-726. $\sqrt[5.6]{0.0018}$ = **0.323**
2-727. $\sqrt[0.67]{0.954}$ = **0.932**

General guides for decimal location

The student should be able to estimate the approximate answer and thereby know on which scale the answer will be found.

The following suggestions are presented so that the student can decide more easily whether the answer is to be larger or smaller than the original quantity.

$$(\text{Number})^{\text{Exponent}} = \text{Answer}$$

1. If the number is larger than 1.00 and the exponent is larger than 1.00, the answer will be greater than the number.
2. If the number is less than 1.00 and the exponent is less than 1.00, the answer will be greater than the number.
3. If the number is less than 1.00 and the exponent is greater than 1.00, the answer will be less than the number.
4. If the number is greater than 1.00 and the exponent is less than 1.00, the answer will be less than the number.

Results that do not fall within the limits of the scales

In many computations the final answer may be larger than 22,026 and hence cannot be read within the limits of the scales. In such cases the original expression must be factored before attempting to use the log-log scales. Several such methods of factoring are explained below.

These methods are for use in finding the powers of numbers. For problems involving roots of numbers convert the problem to one involving the power of a number and then apply the appropriate method.

Example

$$\sqrt{5} = (5)^{1/2} = (5)^{0.5}$$
$$\sqrt[4]{5} = (5)^{1/4} = (5)^{0.25}$$
$$\sqrt[0.5]{5} = (5)^{1/0.5} = (5)^{2}$$

Method 1 Express the number in scientific notation and raise each part to the given power.

Example

$$(35.3)^4 = ?$$
$$(35.3)^4 = (3.53 \times 10)^4$$
$$= (3.53)^4 \times (10)^4$$

Now, using the Lon (LL) scales, and since $(3.53)^4 = 155$, we obtain

$$(35.3)^4 = 155. \times 10^4$$
$$= \mathbf{1.55 \times 10^6} \text{ (Answer)}$$

Method 2 Factor the number which is to be raised to a power and then treat each part separately, as in Method 1.

Example

$$(15)^5 = ?$$
$$(15)^5 = (3 \times 5)^5$$
$$= (3)^5 \times (5)^5$$
$$= (243)(3125)$$
$$= \mathbf{7.59 \times 10^5} \text{ (Answer)}$$

Method 3 Divide the exponent into two or more smaller parts and, using the log-log scales, compute each part separately. A final computation is made using the C and D scales as in Method 1 and Method 2.

Example

$$(2.36)^{15} = ?$$
$$(2.36)^{15} = (2.36)^5 \times (2.36)^5 \times (2.36)^5$$
$$= (73.2)(73.2)(73.2)$$
$$= \mathbf{3.92 \times 10^5} \text{ (Answer)}$$

or

$$(2.36)^{15} = (2.36)^8 \times (2.36)^7$$
$$= (960)(410)$$
$$= \mathbf{3.92 \times 10^5} \text{ (Answer)}$$

$$(2.36)^{15} = (2.36)^{7.5} \times (2.36)^{7.5}$$
$$= (6.20)^2$$
$$= \mathbf{3.92 \times 10^5} \text{ (Answer)}$$

Example

$$(0.000025)^{1.3} = ?$$
$$(0.000025)^{1.3} = (2.5 \times 10^{-5})^{1.3}$$
$$= (2.5)^{1.3} \times (10^{-5})^{1.3}$$
$$= 3.29 \times (10)^{-6.5}$$
$$= (3.29)(10)^{-6}(10)^{-0.5}$$
$$= (3.29)(10)^{-6}\left(\frac{1}{3.16}\right)$$
$$= (3.29)(10)^{-6}(0.316)$$
$$= \mathbf{1.041 \times 10^{-6}} \text{ (Answer)}$$

Method 4 Express the number in scientific notation and then express the power of 10 in logarithmic form.

Example

$$(250)^{3.2} = ?$$
$$(250)^{3.2} = (2.50 \times 10^2)^{3.2} = (2.50)^{3.2}(10)^{6.4}$$

where $(10)^{6.4} = x$ may be expressed as $\log_{10} x = 6.4$ or $x = (2.51)(10)^6$.

Then
$$(2.50)^{3.2}(10)^{6.4} = (1.87 \times 10^1)(2.51 \times 10^6)$$

and
$$(1.87 \times 10^1)(2.51 \times 10^6) = \mathbf{4.71 \times 10^7}$$

Method 5 This method is more suitable for those numbers which have 5, 6, 7, 8, or 9 as the first digit.

Example
$$\begin{aligned}(645)^{13} &= ?\\ (645)^{13} &= (0.645)^{13}(10^3)^{13}\\ &= (0.00334)(10)^{39}\\ &= \mathbf{3.34(10)^{36}} \text{ (Answer)}\end{aligned}$$

Method 6 Factor the exponent such that one part is equivalent to an exact power of ten.

Example
$$(2)^{52} = ?$$

First raise the base (2) to a power such that the answer is an exact power of ten.

$$\begin{aligned}(2)^k &= 10{,}000 = (10)^4\\ k &= 13.29\end{aligned}$$

Also
$$\begin{aligned}(2)^{52} &= (2)^{13.29+13.29+13.29+12.13}\\ &= (10^4)(10)^4(10)^4(2)^{12.13}\\ &= (10^4)^3(2)^{12.13}\\ &= (10)^{12}(4500)\\ &= \mathbf{(4.5)(10)^{15}} \text{ (Answer)}\end{aligned}$$

Example
$$(1.324)(10)^{-9} = (0.815)^m$$

First choose a factor such that an exact power of ten is obtained.

$$(0.815)^{45} = 0.0001 = (10)^{-4}$$

Then
$$\begin{aligned}(1.324)(10)^{-9} &= (0.815)^{45+45+t}\\ &= (0.815)^{45}(0.815)^{45}(0.815)^t\\ &= (10)^{-4}(10)^{-4}(0.815)^t\end{aligned}$$

$$\frac{(1.324)(10)^{-9}}{(10)^{-8}} = (0.815)^t$$

$$\begin{aligned}(1.324(10)^{-1} &= (0.815)^t\\ t &= 9.87\end{aligned}$$

Therefore $(1.324)(10)^{-9} = (0.815)^{45+45+9.87}$
$(1.324)(10)^{-9} = (0.815)^{99.87}$

and $m = \mathbf{99.9}$ (Answer)

Methods 1 and 6 are generally preferred over the other methods because they usually make greater accuracy possible in the final answer.

Finding the natural logarithm of a number

The natural base for logarithms is $\epsilon(2.71828-)$. The logarithm of any number (to the base ϵ) may be found as follows:

For number — For numbers greater than 1.00

$$\log_\epsilon X = A$$

1. Locate the number X on the Ln0, Ln1 (LL_1), Ln2 (LL_2), or Ln3 (LL_3) scale.
2. Read the logarithm of the number under the hairline on the D scale.

Location of Decimal Point

If the number X is on	Decimal point in the answer is
Ln3 or LL_3	x.xxx
Ln2 or LL_2	0.xxx
Ln1 or LL_1	0.0xxx
Ln0	0.00xx

Examples

$\log_\epsilon 62 = 4.13$
$\log_\epsilon 1.271 = 0.240$
$\log_\epsilon 1.026 = 0.0257$

For numbers less than 1.00

$$\log_\epsilon X = A$$

1. Locate the number X on the Ln-0, Ln-1 (LL_{01}), Ln-2 (LL_{02}), or Ln-3 (LL_{03}) scales.
2. Read the logarithm (to the base ϵ) of the number A directly above X on the D scale.

Location of Decimal Point

If the number X is on	Decimal point in the answer is
Ln-3 or LL_{03}	–x.xxx
Ln-2 or LL_{02}	–0.xxx
Ln-1 or LL_{01}	–0.0xxx
Ln-0	–0.00xx

3. The logarithm (to the base ϵ) of all numbers less than 1.000 is a negative number.

Examples

$$\log_\epsilon 0.0045 = -5.40$$
$$\log_\epsilon 0.745 = -0.294$$
$$\log_\epsilon 0.954 = -0.0471$$

Problems

2-728. $(2.89)^{6}$
2-729. $(4.11)^{5.2}$
2-730. $(19.01)^{1.6}$
2-731. $(1.185)^{2.7}$
2-732. $(1.033)^{5.8}$
2-733. $(1.0134)^{25}$
2-734. $(3.95)^{0.65}$
2-735. $(8.46)^{0.134}$
2-736. $(81.2)^{0.118}$
2-737. $(7850.)^{0.0775}$
2-738. $(1.399)^{0.883}$
2-739. $(10.06)^{0.0621}$
2-740. $(0.569)^{4}$
2-741. $(0.157)^{8}$
2-742. $(0.985)^{1.568}$
2-743. $(0.318)^{4.65}$
2-744. $(0.078)^{0.458}$
2-745. $(17.91)^{0.012}$

2-746. $(4780.)^{0.913}$
2-747. $(253.)^{0.269}$
2-748. $(0.428)^{0.559}$
2-749. $(4.08)^{24}$
2-750. $(3.91)^{20}$
2-751. $(8.45)^{16}$
2-752. $(7.77)^{42}$
2-753. $(16.89)^{1.402}$
2-754. $(87.8)^{8}$
2-755. $(0.1164)^{0.33}$
2-756. $(0.779)^{0.43}$
2-757. $(867.)^{6}$
2-758. $(91.05)^{14}$
2-759. $(0.775)^{0.0259}$
2-760. $\sqrt[6]{8.69}$
2-761. $\sqrt[5]{1.094}$
2-762. $\sqrt[1.3]{8.74}$

2-763. $\sqrt[0.6]{19.77}$
2-764. $\sqrt[18]{54.8}$
2-765. $\sqrt[7]{1.004}$
2-766. $\sqrt[1.95]{0.642}$
2-767. $\sqrt[14]{0.1438}$
2-768. $\sqrt[3.6]{0.952}$
2-769. $\sqrt[2.4]{0.469}$
2-770. $\sqrt[1.7]{0.1975}$
2-771. $\sqrt[0.55]{0.2218}$
2-772. $\sqrt[0.46]{16{,}430}$
2-773. $\sqrt[0.133]{507.}$
2-774. $\sqrt[0.57]{0.964}$
2-775. $\sqrt[5.09]{6.49}$
2-776. $\sqrt[13.6]{0.1574}$
2-777. $\sqrt[2.09]{0.1268}$

Solve for *X*.

2-778. $X = (43.8)^{6.4}$
2-779. $X = (1.853)^{0.447}$
2-780. $(31.77)^{x} = 1.164$
2-781. $(2.388)^{3x} = 3.066$
2-782. $(1.064)^{0.2x} = 4.99$
2-783. $(X)^{5.8} = 8.57$
2-784. $(4.92)^{0.66x} = 24.1$
2-785. $(0.899)^{4.7x} = (1.552)(10)^{-8}$
2-786. $(0.1135)^{0.77x} = 0.775$
2-787. $(11.774)^{8.31x} = 12.88$
2-788. $(18.73)^{6.4x} = 8688.$
2-789. $(34.86)^{1.117x} = 9.44$
2-790. $(0.631)^{0.64x} = 0.318$
2-791. $(0.1299)^{0.68x} = 0.443$
2-792. $(15.84)^{x} = 4.87$
2-793. $(0.679)^{x} = 0.337$

2-794. $(1.461)^{19.66x} = 9.07$
2-795. $(0.766)^{5.8x} = 0.239$
2-796. $(X)^{7.99} = 0.775$
2-797. $(X)^{0.175} = 8.53$
2-798. $(X)^{3.33} = 1.055$
2-799. $(X)^{0.871} = 0.1557$
2-800. $(X)^{4.77} = 1.088$
2-801. $(X)^{0.771} = 0.0521$

2-802. $(4.51)^{0.199} = \dfrac{X}{3}$

Solve for the natural logarithms of the following numbers

2-803. 15.77
2-804. 19,850
2-805. 0.7789
2-806. 0.1845
2-807. 1.896
2-808. 56.87
2-809. 13.09
2-810. 33.4
2-811. 8.09
2-812. 1.571
2-813. 0.1345
2-814. 0.915
2-815. 0.001233
2-816. 13,890.
2-817. 2.066
2-818. 1.3157
2-819. 1.0047
2-820. 89.78
2-821. 0.664
2-822. 0.459
2-823. 0.1175
2-824. 1.9974
2-825. 0.9974
2-826. 0.2378
2-827. 0.01663

Review problems

Solve by general slide rule methods.

2-828. (51)(9)
2-829. (426)(51)
2-830. (6.03)(5.16)
2-831. (561)(4956)
2-832. (43.2)(0.617)
2-833. (6617)(0.00155)
2-834. (99.043)(3.091)
2-835. (0.0617)(0.4417)
2-836. $(1.035)(2.31 \times 10^5)$
2-837. $(79.81 \times 10^{-4})(0.617)$
2-838. $(516 \times 10^{-8})(0.391 \times 10^{-2})$
2-839. (51)(97)(32)
2-840. (52.3)(759.3)
2-841. (716.5)(0.03166)
2-842. (11.65)(−0.9213)
2-843. (76.2)(−31.45)
2-844. (−0.6175)(−12,391)
2-845. $\frac{(-759.6)}{(0.6175)}$
2-846. $\frac{(-19.96)}{(3346)}$
2-847. $\frac{(-1.0366)}{(29.31)}$
2-848. $\frac{(7575)}{(695.2)}$
2-849. $\frac{(-516.6)}{(0.06052)}$
2-850. (116.5)(4619)(0.317)
2-851. (210.9)(151.3)(7716)

2-852. $(706.5)(1.695 \times 10^{-6})(0.006695)$
2-853. $(1033)(7.339 \times 10^{-6})(0.0317 \times 10^{-3})$
2-854. $(4.017 \times 10^{-8})(0.0991)(0.1756)$
2-855. $(5.576)(0.0917)(1.669 \times 10^4)$
2-856. (6.991)(0.75)(0.993)(4.217)
2-857. $(56.88)(0.971 \times 10^{-5})$
2-858. (59.17)(0.3617)(0.5916)(0.00552)
2-859. (5.691)(0.3316)(0.991)(0.00554)(0.1712)
2-860. (6.523)(71.22)(4.091)(591)(600)(0.1332)
2-861. $(43.06)(0.2361)(0.905 \times 10^{-4})(3.617 \times 10^{-3})$

2-862. $(1917)^{2.16}$
2-863. $(4.216)^{1.517}$
2-864. $(2.571)^{2.91}$
2-865. $(0.3177)^{2.06}$
2-866. $\sqrt[5]{26.31}$
2-867. $\sqrt[3]{0.03175}$

2-868. $\sqrt{116.75}$

2-869. $\sqrt[3]{0.6177}$

2-870. $\sqrt{3167}$

2-871. $(179 \times 10^3)(0.3165)$

2-872. $(5033 \times 10^{-4})(0.9116)$

2-873. $(0.06105)(77.165)$

2-874. $(\sqrt{216})(34)(\pi)^2$

2-875. $(\sqrt{819})(107)(\sqrt{\pi})$

2-876. $\dfrac{(\sqrt{616})(6.767)}{(\sqrt{39.6})}$

2-877. $\dfrac{(1045)}{(X)} = \dfrac{(0.0278)}{(0.0798)}$

2-878. $\dfrac{(1.486)}{(33)} = \dfrac{(0.37)(X)}{467}$

2-879. $(816) = \dfrac{(244)(2\pi)}{(0.049)(X)}$

2-880. $(0.0036)(\sin 49.8°)$

2-881. $\dfrac{(20.5)^2(7.49)(\sin 49°)}{(30.5)(0.0987)}$

2-882. $\sqrt{\dfrac{(38)^2(6.71)^2}{\pi}}$

2-883. $(7.61)(\sqrt[3]{7.61})(\pi)$

2-884. $\dfrac{(13.1)(\sin 3.12°)}{(\tan 41.9°)}$

2-885. $\dfrac{2}{3} = \dfrac{(X)(\pi)}{8.37}$

2-886. $\dfrac{(9616)}{X} = \dfrac{(3.1416)}{(0.0142)}$

2-887. $(\sqrt[3]{64.9})(2.1 \times 10^3)$

2-888. $(4 \times 10^6)(0.007) = (X)(10{,}980)$

2-889. $Y = \left(\frac{1}{4}\right)\left(\frac{16}{6}\right)\left(\frac{1}{17}\right)$

2-890. $\dfrac{X}{\pi} = \dfrac{(\sqrt{46.2})(3.14)^2}{(\sin 3.7°)}$

2-891. $\dfrac{(3.98)(X)}{(1.07)(38)} = \dfrac{(3 \times 10^6)}{(17{,}680)}$

2-892. $\dfrac{(\sqrt[3]{986})}{X} = \dfrac{(14)}{(1/116)}$

2-893. $\dfrac{(X)^2}{(9.2)} = \dfrac{(18.17)(3.4)}{(166)}$

2-894. $\dfrac{(3.6)}{(X)^2} = \dfrac{(9.6 \times 10^2)}{(67.4)} = \dfrac{(Y)^{1/2}}{(64)}$

2-895. $\dfrac{(X)^{1/2}}{(31.1)} = \dfrac{(\sqrt{196})(189.1)}{4/76}$

2-896. $\dfrac{(96.5)}{(3.9)} = \dfrac{X}{(\sin 46.6°)} = \dfrac{(Y)^2}{(3.14 \times 10^{-2})}$

2-897. $\dfrac{(X)^2}{Y} = \dfrac{(67.3)^2(Y)}{(96.61)} = \dfrac{(497.1)}{\tan 75°}$

2-898. $\dfrac{(3.7)(4.9)}{X} = \dfrac{(46.7)}{564}$

Solve by general slide rule methods.

2-899. $\dfrac{Y}{(28)} = \dfrac{(3.2)}{(4/118)}$

2-900. $\dfrac{Y}{42} = \dfrac{39.1}{(1/45)}$

2-901. $(37.3)(X)(46.6) = (175(\pi)$

2-902. $(\sqrt{256})(3) = (X)(197.6)$

2-903. $\dfrac{(54.6)(\tan 10.6°)}{(\sqrt{0.0967})(8.1 \times 10^3)}$

2-904. $\dfrac{\sqrt[3]{(15.1)^2}(31.4)^2}{(\sin \text{arc} \cos 0.617)}$

2-905. $\dfrac{(0.954)(0.06 \times 10^3)}{(\tan 59°)^{1/2}(6.5)^2}$

2-906. $\dfrac{\sqrt[3]{(15.6)^2}(0.9618)}{(0.08173)(61{,}508)(2\pi)}$

2-907. $\dfrac{(68)(765)(391)(0.0093 \times 10^3)}{(571)^2(\sqrt[3]{(64)})}$

2-908. $\dfrac{(\cos 11.5°)(\sqrt{6.87})}{(0.00081)(7.7 \times 10^4)}$

2-909. $\dfrac{\sqrt[4]{(1.71)^5}(6.87)}{(\tan 53°)(5.1)^2}$

2-910. $\dfrac{(0.000817)(\tan 81°)}{(0.00763)(\sin 81°)}$

2-911. $(273)^{1/2}(46.9)(\cos 61°)(\pi^3)$

2-912. $\dfrac{(\sin \text{arc} \tan 3.17)(71.7)}{(\sqrt{89.6})(\sqrt[4]{(76.5)^2})}$

2-913. $\dfrac{(\sqrt{(16)^3})(\log_{10} 100)}{(6.71 \times 10^{-1})(3.71)^3}$

2-914. $\dfrac{(6.93)(\sin \cos^{-1} 0.98)}{(0.937)^2(39.6)}$

2-915. $\dfrac{(\sqrt{91.68})(\sqrt[3]{65.9})}{(\tan 68.7°)(0.671)^2}$

2-916. $\dfrac{(4.5)^4(\sqrt{(98.71)})(\sin 56.4°)}{(0.09 \times 10)(38.6)^{3/2}}$

2-917. $\dfrac{(\sqrt{285})(\cos 36.6°)(1.64)^2}{(67.1 \times 10^{-1})(5780)}$

2-918. $\dfrac{(\tan \sin^{-1} 0.87)(61.7)}{(5.64)^{0.98}(3.65)^2}$

2-919. $\dfrac{(3174)(\tan 64°)}{(81.6)^2(\sqrt[3]{18})}$

2-920. $\dfrac{(44.6)(0.09 \times 10^3)(\sin 80.9°)}{(\sqrt[3]{96.7})(51.6)^2}$

2-921. $\dfrac{(\tan 50.6°)(3.4)^2}{(\sqrt{9681})(171)}$

2-922. $\dfrac{(296)(0.197 \times 10^5)}{\sqrt[4]{(76.1)}(\sin 49.6°)}$

2-923. $\dfrac{(\sin 22.6°)(9.918)}{(\tan 31.6°)(98.71)}$

2-924. $\dfrac{(68.7 \times 10^2)(\tan 56.1°)}{(96.7)^{0.86}(18{,}614)}$

2-925. $\dfrac{(0.0098)(\sin 17.6°)\sqrt{(0.186)}}{(41.6)^2(689.0)}$

2-926. $\dfrac{(\tan 19.8°)^2(6.71 \times 10^3)}{(1{,}876)(\sqrt[4]{59})}$

2-927. $\dfrac{(\sqrt{\sin 40°})(17)^2(4\pi^2)}{(0.643)(\tan 60°)}$

Hyperbolic functions on the slide rules

Hyperbolic functions are useful in several mathematical applications such as the variation of electrical current and voltage with distance in the calculation of transmission of electrical power. Several manufacturers of slide rules make special scales from which hyperbolic functions can be read directly. However, it is possible to obtain numerical values for hyperbolic functions using conventional scales by making use of the relations:

$$\frac{\epsilon^x - \epsilon^{-x}}{2} = \text{hyperbolic sine } x \ (\sinh x)$$

$$\frac{\epsilon^x + \epsilon^{-x}}{2} = \text{hyperbolic cosine } x (\cosh x)$$

$$\frac{\epsilon^{2x} - 1}{\epsilon^{2x} + 1} = \text{hyperbolic tangent } x \ (\tanh x)$$

Reading hyperbolic scales

Most slide rules that have hyperbolic scales have the scales marked as Sh and Th. Slide rules manufactured by Pickett identify the hyperbolic sine scales as *upper* and *lower* and the values of sinh x are read on the C scale. Keuffel & Esser identify the hyperbolic sine scales as Sh 1 and Sh 2 and values of sinh x are read on the D scale. Except for these minor differences, reading hyperbolic functions on slide rules made by either company is essentially the same.

Hyperbolic sines In order to read hyperbolic sine functions on the slide rule, set the value sinh x on one of the Sh scales and read the value of the function on either the C scale or the D scale under the hairline.

Example Find sinh 0.38.

Solution Locate 0.38 on the upper Sh scale or on the Sh 1 scale and read 0.389 on the C or D scale.

Example Find sinh 1.88.

Solution Using the method above read sinh 1.88 = 3.20. Note that the value 1.88 is located on the lower Sh scale (Sh 2 scale) and 3.20 is read on the C scale (D scale).

The decimal point can be determined readily by noting that numbers corresponding to function values on the upper Sh (Sh 1) scale lie between 0.1 and 1.0, and numbers corresponding to function values on the lower Sh (Sh 2) scale lie between 1.0 and 10.0.

Hyperbolic tangents Hyperbolic tangents can be read by locating the value of the tangent function on the Th scale and reading the number on the C or D scale under the hairline.

Example tanh 0.206 = **0.1990**

Example tanh 1.33 = **0.870**

Hyperbolic Cosines Most slide rules do not have a hyperbolic cosine scale. Values for the hyperbolic cosine can be determined by use of the relation:

$$\cosh x = \frac{\sinh x}{\tanh x}$$

In finding values for cosh x using the Pickett rule, first set the slide so the indexes coincide. Locate the hairline over the value of x on the appropriate Sh scale. Move the slide until the value of x on the Th scale is under the hairline and cosh x can be read on the D scale at the C index.

Example cosh 0.482 = **1.118**

Example cosh 1.08 = **1.642**

For the Keuffel & Esser Vector slide rule, this procedure can be followed. Set an index of the slide on the value of x on the Th scale. Set the hairline on the value of x on either Sh 1 or Sh 2, depending on its amount. Read the value of cosh x on the C scale.

Example cosh 0.305 = **1.046**

Example cosh 1.181 = **2.31**

When the value of cosh x is given and it is desired to find x, use can be made of the relation

Example $$\cosh^2 x - \sinh^2 x = 1$$

Find the value of x when $\cosh x = 2.1$

Solution $$\sinh x = \sqrt{\cosh^2 x - 1}$$

Substituting:

$$\sinh x = \sqrt{(2.1)^2 - 1}$$
$$\sinh x = \sqrt{3.41}$$

and

$$\sinh x = 1.85$$

Set 1.85 on the C (D) scale and read the value of x on the lower Sh (Sh 2) scale. The lower scale is used because $\sinh x$ is greater than 1.

Then $$x = \mathbf{1.372}$$

Approximations for large and small values of x When the value of x is more than 3, it can be shown that the value of $\sinh x$ and $\cosh x$ is approximately the same as $\frac{\epsilon^x}{2}$.

Example $$\sinh 4.2 = ?$$

$$\frac{\epsilon^{4.2}}{2} = 33.5$$

$$\sinh 4.2 \cong \mathbf{33.5}$$

Also for large values of x, $\tanh x$ is approximately 1.0.

Example $$\tanh 3.7 = ?$$

Solution $$\tanh 3.7 = \frac{\epsilon^{(2)(3.7)} - 1}{\epsilon^{(2)(3.7)} + 1}$$

$$= \frac{1650 - 1}{1650 + 1}$$

$$\tanh 3.7 \cong \mathbf{1.0}$$

When x has values below 0.1, it can be shown that $\sinh x$ and $\tanh x$ are approximately the same as x, and $\cosh x$ is approximately 1.0.

Example

$$\sinh 0.052 \cong \mathbf{0.052}$$
$$\tanh 0.037 \cong \mathbf{0.037}$$
$$\cosh 0.028 \cong \mathbf{1.00}$$

Other hyperbolic functions While not often needed, other hyperbolic functions can be obtained by using the following defining expressions:

$$\coth x = \frac{1}{\tanh x}$$

$$\operatorname{sech} x = \frac{1}{\cosh x}$$

$$\operatorname{csch} x = \frac{1}{\sinh x}$$

Problems on hyperbolic functions

2-928. Find the values of $\sinh x$ for the following values of x: (*a*) 0.12, (*b*) 1.07, (*c*) 1.91, (*d*) 2.30, (*e*) 3.11, (*f*) 4.26, (*g*) 5.00

2-929. Find the values of x for the following values of $\sinh x$: (*a*) 0.1304, (*b*) 0.956, (*c*) 1.62, (*d*) 4.10, (*e*) 8.70, (*f*) 19.42, (*g*) 41.96

2-930. Find the values of $\cosh x$ for the following values of x: (*a*) 0.28, (*b*) 1.03, (*c*) 1.98, (*d*) 2.37, (*e*) 3.56, (*f*) 4.04, (*g*) 5.00

2-931. Find the values of x for the following values of $\cosh x$: (*a*) 1.204, (*b*) 1.374, (*c*) 2.31, (*d*) 5.29, (*e*) 8.50, (*f*) 21.7, (*g*) 52.3

2-932. Find the values of $\tanh x$ for the following values of x: (*a*) 0.16, (*b*) 0.55, (*c*) 1.14, (*d*) 1.94, (*e*) 2.34, (*f*) 2.74, (*g*) 5.00

2-933. Find the values of x corresponding to the following values of $\tanh x$: (*a*) 0.1781, (*b*) 0.354, (*c*) 0.585, (*d*) 0.811, (*e*) 0.881, (*f*) 0.980, (*g*) 0.990

Slide rule solution of complex numbers

A complex number, which consists of a real part and an imaginary part, is often used to describe a vector quantity. By definition, a vector quantity, frequently referred to as a *phasor* in electrical engineering, has both magnitude and direction. For example, the expression $3 + j4$ will describe a vector which is $\sqrt{3^2 + 4^2}$ units long and makes an angle arc $\tan\frac{4}{3}$ with an x axis. For a more complete discussion on complex number theory, refer to a text on basic algebra.

The symbol i or the symbol j is customarily used to represent the quantity $\sqrt{-1}$. In the discussion in this section the symbol $j = \sqrt{-1}$ will be used.

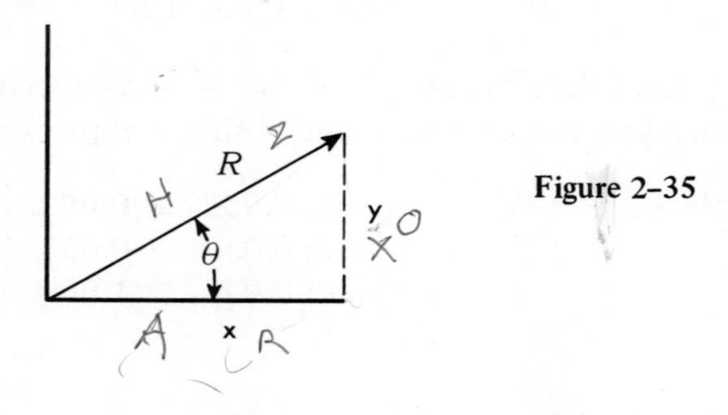

Figure 2-35

If we let the scalar length of a vector be designed as R, as shown in Figure 2–35, then we can write $R\epsilon^{j\theta} = x + jy$ in polar form as $R\underline{/\theta}$. This expression $R\underline{/\theta}$ is a shortened form of $R\epsilon^{j\theta}$ which is obtained from the identity

$$R\epsilon^{j\theta} = R \cos \theta + jR \sin \theta.$$

Complex numbers on the slide rule From trigonometric relations for a right triangle, we can show that $\tan \theta = \frac{y}{x}$, $R = \frac{y}{\sin\theta}$, and $R = \frac{x}{\cos\theta}$. We can use these relations to solve complex number problems on the slide rule. Take, for example, the complex number $3 + j4$ and let it be required to find $R\underline{/\theta}$.

The following method will give the solution to this problem on most types of slide rules:

1. Locate the larger of the two numbers on the D scale and set an index of the C scale at this number. Locate the smaller of the two numbers on the D scale using the hairline, and read the angle θ on the T scale under the hairline. If y is smaller than x, θ is less than 45°, and if y is larger than x, θ is larger than 45°.
2. Next move the slide until the angle θ on the S scale is in line with the smaller of the two numbers. Read R on the D scale at the index of the C scale.

Example Express $3 + j4$ in polar form.

Solution Set the right index of the C scale at 4 on the D scale.

Move the hairline to 3 on the D scale and read $\theta = 53.1°$ on the T scale. Note that the y value is larger than the x value; thus the angle is larger than 45°.

Without moving the hairline, move the slide until 53.1° on the S scale (reading angles to the left) is under the hairline.

Read 5 at the right index of the C scale.

The solution is $3 + j4 = \underline{5\underline{/53.1°}}$

This method can be performed on most types of rules, requiring the minimum number of manipulations of the rule. It also can be applied readily to the solution of most problems involving right triangles.

When any of the complex numbers have a minus sign, the slide rule operation to solve the problem is the same as though the sign of the numbers were positive. The angles usually are determined by inspection using trigonometric relations. The following general rules apply:

If the expression has the form $+x + jy$, θ is in the first quadrant.

For $-x + jy$, θ is in the second quadrant
For $-x + jy$, θ is in the third quadrant
For $+x + jy$, θ is in the fourth quadrant

Example Express $-7.1 + j3.8$ in polar form.

Solution Set the right index of the C scale at 7.1 on the D scale, and read on the T scale $\theta = 28.3°$ at 3.8 on the D scale.

Move the slide so that 28.3° on the S scale is over 3.8 and read $R = 8.03$ at the C index.

By inspection, the angle is in the second quadrant and the total angle is $180° - 28.3° = 151.7°$.

Therefore, the polar form is $\mathbf{8.03\angle 151.7°}$.

Example Express $4 - j3$ in polar form.

Solution The angle is read as 36.9° and is in the fourth quadrant. The total angle is $360° - 36.9° = 323.1°$.

The polar form is $\mathbf{5\angle 323.1°}$.

If the polar form is given, the rectangular form can be obtained by multiplying the value of R by the appropriate sine and cosine value. A rapid method of finding the quantities is to use the previously described slide rule manipulation in reverse.

Example Express $3.3\angle 28°$ in rectangular form.

Solution Set the right index of the C scale at 3.3 on the D scale and read 1.55 on D under 28° on the sine scale.

Move the slide until 28° on the T scale is over 1.55 on the D scale.

Read 2.915 on the D scale at the right index. Since the angle is less than 45°, the imaginary part of the complex number is the smaller of the two. Therefore,

$$3.3\angle 28° = \mathbf{2.915 + j1.55}$$

If the polar angle is larger than 45°, angles on the T scale and S scale are read to the left, and the real part of the complex number is read first. The real part of the number will be the smaller of the two parts.

Example $179\angle 66° = \mathbf{72.9 + j163.5}$

For angles not in the first quadrant, obtain the angle of the vector with respect to the x axis and treat the solution as outlined above. By inspection, affix the proper signs to the real and imaginary parts after obtaining their values. A sketch will help greatly in this process.

Conversion for small angles If the ratio of the x value and y value in the complex number is greater than 10, the angle can be found on the ST scale. The real value is approximately equal to the value of R.

Example $35 + j1.5 = R\underline{/\theta}$

Solution Set the C index at 35 on the D scale and read $\theta = 2.45°$ on the ST scale.

Then $R\underline{/\theta} \cong \mathbf{35\underline{/2.45°}}$.

Example $0.0075\underline{/4.1°} = x + jy$

Solution Set the C index at 0.075 on the D scale. Read 0.00536 on D under 4.1° on ST.

Then $x + jy \cong \mathbf{0.075 + j0.00536}$.

Conversion for angles near 90° For angles between 84.27° and 90°, the ratio of x to y will be 10 or greater and the imaginary part of the complex number is approximately equal to the value of R. The angle can be read on the ST scale after subtracting it from 90°.

Example $18\underline{/88°} = x + jy$

Solution Set the left index of C at 18 on the D scale. Read 0.6 on D under 2° on the ST scale.

Then $x + jy \cong \mathbf{0.6 + j18}$.

Remember that for very large and very small angles, the ratio of x and y will be 10 or greater, and either the real part or the imaginary part of the complex number will be approximately equal to the value of R.

Applications of complex numbers In solving problems involving complex numbers, addition and subtraction of complex numbers are more easily performed if the numbers are expressed in rectangular form. In this form, the respective real parts and imaginary parts can be added or subtracted directly. However, to multiply or divide complex numbers, it is more convenient to express them in polar form and solve by multiplying or dividing the vector magnitude, and adding or subtracting the angular magnitude.

Examples

$$(a + jb) + (c + jd) = (\mathbf{a + c}) + \mathbf{j(b + d)} \quad \text{(Addition)}$$

$$(a + jb) - (c + jd) = (\mathbf{a - c}) + \mathbf{j(b - d)} \quad \text{(Subtraction)}$$

$$(a/\theta_1)(b/\theta_2) = (\mathbf{a})(\mathbf{b})\underline{/\theta_1 + \theta_2} \quad \text{(Multiplication)}$$

$$\frac{a\underline{/\theta_1}}{b\underline{/\theta_2}} = \frac{\mathbf{a}}{\mathbf{b}}\,\underline{/\theta_1 - \theta_2} \quad \text{(Division)}$$

From the examples above, we can see that the ability to perform rapid conversions from polar form to rectangular form or vice versa will be helpful in solving problems involving complex numbers.

Problems on complex numbers

2-934. Express in polar form: (*a*) $8 + j3$, (*b*) $2 + j6$, (*c*) $1 + j4$, (*d*) $5 + j5$

2-935. Express in rectangular form: (*a*) $6.2\angle 39°$, (*b*) $3.6\angle 48°$, (*c*) $9.2\angle 21.4°$, (*d*) $2.7\angle 71°$

2-936. Express in polar form: (*a*) $-8.9 + j4.2$, (*b*) $-16.8 + j9.3$, (*c*) $-5.3 + j2.1$, (*d*) $-18.4 + j3.3$

2-937. Express in rectangular form: (*a*) $9.7\angle 118°$, (*b*) $115\angle 137°$, (*c*) $2.09\angle 160°$, (*d*) $5.72\angle 110°$

2-938. Express in polar form: (*a*) $-7.3 - j6.1$, (*b*) $-4.4 - j8.2$, (*c*) $-8.8 - j2.5$, (*d*) $-1.053 - j5.13$

2-939. Express in rectangular form: (*a*) $81.3\angle 200°$, (*b*) $62.1\angle 253°$, (*c*) $1059\angle 197°$, (*d*) $0.912\angle 231°$

2-940. Express in polar form: (*a*) $160.5 - j147$, (*b*) $89.3 - j46.2$, (*c*) $0.0062 - j0.0051$, (*d*) $3.07 - j1.954$

2-941. Express in rectangular form: (*a*) $557\angle 297°$, (*b*) $6.03\angle 327°$, (*c*) $0.9772\angle 344°$, (*d*) $19{,}750\angle 300°$

2-942. Express in polar form: (*a*) $15.61 + j7.09$, (*b*) $-14.9 - j61.7$, (*c*) $0.617 - j0.992$, (*d*) $-41.2 + j75.3$

2-943. Express in rectangular form: (*a*) $1.075\angle 29.1°$, (*b*) $10.75\angle 136°$, (*c*) $107.5\angle 253°$, (*d*) $1075\angle 322°$

3

the electronic hand calculator

As a calculation device in problem solving the electronic hand or "pocket" calculator has no peer. It is vastly superior to the slide rule in many respects—convenience, speed, accuracy, versatility, and capability. As the unit cost of such calculators continues to decrease, so will their utilization by engineers, scientists, technicians, and students continue to increase. It is probable that within a few years slide rules will be replaced almost completely by electronic hand calculators. Several of the most popular hand calculators will be discussed in this chapter. In each case particular attention is given to arithmetic operations, the roots and powers of numbers, and trigonometric functions.

When one is working at a desk, the calculator should be operated with the non-writing hand. For example, right-handed persons should operate the calculator with the left hand.

CALCULATOR LOGIC

Electronic hand calculators are designed to perform their functions using either (1) Lukasciewicz (so-called "reverse-Polish") logic with operational stack, or (2) algebraic logic. There are advantages and disadvantages for each system. The fundamental differences in the manual operation for each type of logic can be illustrated by comparing the operating procedures required to solve the three simple problems shown on top of the following page.

	Keyboard Entries Required for the Solution	
Problem	*Lukasciewicz Logic Calculators*	*Algebraic Logic Calculators*
(2 × 3) + (4 × 5) = ?	[2] [↑] [3] [×] [4] [↑] [5] [×] [+]	[2] [×] [3] [+] [4] [×] [5] [=]
(2 + 3) × (4 + 5) = ?	[2] [↑] [3] [+] [4] [↑] [5] [+] [×]	[2] [+] [3] [=] [STO] [4] [+] [5] [=] [×] [RCL] [=]
(2 + 3) ÷ (4 + 5) = ?	[2] [↑] [3] [+] [4] [↑] [5] [+] [÷]	[2] [+] [3] [=] [STO] [4] [+] [5] [=] [÷] [RCL] [$x \gtrless y$] [=]

To the casual or inexperienced observer who might infrequently pick up an electronic hand calculator, algebraic logic is definitely preferred. In this case each key stroke follows in the same sequence that was previously learned in elementary algebra. Because of this feature, the keyboard routine required to use a calculator with algebraic logic is very easy to master. However, as shown in the examples above, the number of key strokes or entries required to solve a particular problem (when using algebraic logic) is usually more than is required when Lukasciewicz logic and operational stack is employed. For involved problems this can become a very important factor. Calculators using Lukasciewicz logic also automatically display intermediate answers. This allows for the inspection of all sub-steps of a calculation and further enhances operator confidence and the dependability of obtaining accurate answers. Also, in some instances it is desirable to be able to record intermediate answers for future use.

Since most electronic hand calculators are individually owned, the simplicity of using a machine designed to operate with algebraic logic becomes less advantageous. Calculators designed to use Lukasciewicz logic are in fact quite simple to use once one learns the specific operating procedures and techniques. In addition, this particular routine must be mastered only once in the user's lifetime. Once the use of this logic system is mastered, every problem is solved in exactly the same way, and there is no need to restructure a problem to make it conform to machine logic, as would be the case with a calculator using algebraic logic and that does not have parentheses keys. If provision is made for nesting of parenthetical expressions, then the two systems become comparable as far as utility of operation is concerned. Lukasciewicz logic and operational stack type calculators also provide the capability to review all stored data in the machine by depressing either the Roll Down Stack [R↓] key or the Register Exchange [$x \gtrless y$] key. This feature is generally not available in algebraic logic type calculators.

Illustration 3-1. The Hewlett-Packard HP-45 Calculator. (Courtesy of Hewlett-Packard.)

THE HEWLETT PACKARD HP-35 AND HP-45

One of the most sophisticated and powerful electronic hand calculators on the market today is the HP-45. Its accuracy exceeds the precision to which most of the physical constants of the universe are known. For example, it will handle numbers as large as 10^{99} or as small as 10^{-99}. It automatically places the decimal point and allows 20 different options for rounding the display to provide flexibility and convenience in interpreting results. Provision is made to handle transcendental functions, such as logarithms (to find the roots and powers of numbers), sines, and cosines; polar/rectangular coordinate conversions for handling complex arithmetic; vectors; selective operating modes; and multiple storage registers. Additionally, constants for π and ϵ are provided—as well as three metric/English unit constants for conversions between centimeters/inches, kilograms/pounds, and liters/gallons. In addition statistical capabilities for calculating the mean (arithmetic average) and standard deviation are incorporated. This calculator has four temporary memory registers (called operational stack), a "Last x" register, and nine registers for user data storage.

The HP–35, although not as elaborate a machine as the HP–45, is capable of executing the primary arithmetic and transcendental functions found in engineering and scientific calculations. It has four registers (or memories) which can be used to store preliminary calculations that may be desired for recall at some later time. Both the HP–35 and HP–45 perform the fundamental arithmetic operations in a similar manner—although there is some variation in the keyboards. A desirable feature of the Hewlett Packard

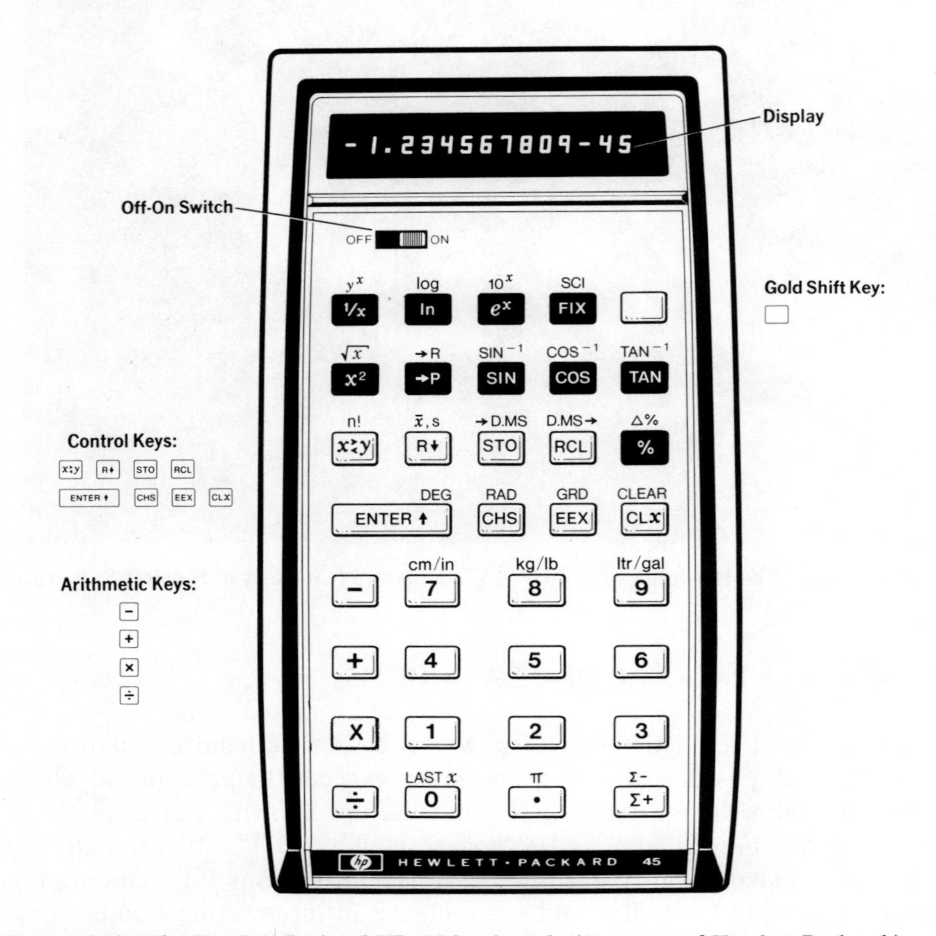

Figure 3–1. The Hewlett-Packard HP–45 keyboard. (Courtesy of Hewlett-Packard.)

calculators is the reliability and touch sensitivity of the numeric and function keys. However, their general operation for evaluating mathematical expressions does differ somewhat from the calculators made by other manufacturers (specifically those described in this chapter) because Hewlett Packard machines use an operational stack and Lukasciewicz notation.

In general, calculator solutions for Hewlett Packard calculators take the same form. Starting at the left side of the equation:

(a) key in the first number and depress the [ENTER ↑] key.

(b) key in the second number and depress the arithmetic operation key needed to calculate the first intermediate answer.

(c) key in the third number and depress the [ENTER ↑] key.

(d) key in the fourth number and depress the arithmetic key needed to calculate the second intermediate answer.

(e) depress the arithmetic operation key needed to calculate the final answer.

Examples $(2 \times 3) + (4 \times 5) = 26$
$(2 + 3) \times (4 + 5) = 45$
$(2 + 3) \div (4 + 5) = 0.56$

For brevity, specific operational instructions are given here only for the HP–45 calculator.

HP–45 Keyboard **Summary of Most Used Functions**

Item	*Name*	*Description of Function*
OFF [switch] ON	Power Switch	The switch should be turned off when the calculator is not in use. All internal storage of calculations within the machine are erased when the power switch is moved to the "OFF" position. Depressing the key slightly before moving it will diminish its wear.
[0] to [9]	Numeric Key Set	Each digit must be keyed in sequence, as desired, by depressing the appropriate number keys. Digits keyed in this manner are stored initially in the X register—the bottom, or display, register.
[•]	Decimal Point	The decimal point symbol must be depressed in the sequential position in which it occurs. Otherwise the calculator register will place a decimal to the right of the last numeric key pressed. Example: 314.32 would be keyed as [3] [1] [4] [•] [3] [2] .

HP-45 Keyboard		Summary of Most Used Functions
Item	*Name*	*Description of Function*
▬	Display Register	The display shows the contents of the X register—the bottom register.
ENTER ↑	Copy x into y	Depress the ENTER ↑ key to terminate the number string being entered into the X register. Depressing this key will also move the displayed value of the X register into a second internal register (a place that holds numbers) which is called the Y register.
CLx	Clear X	To erase the value shown on the display register (X register), depress the Clear x CLx key.
■	Gold Shift Key	The Gold Key ■ should be depressed prior to actuating alternate functions.
⌜ ⌟	Alternate Functions	*Alternate functions* are printed directly above their keyed locations. They are indicated like this, ■ $\sqrt{x}$, in this text.
CLEAR	Clear Calculator	Depressing the Gold Key ■ and then the *alternate function* Clear Calculator CLEAR key will erase everything in the calculator, except for certain data storage registers.
+	Add	To add a number to the value appearing in the display register, terminate the number string by pressing ENTER ↑ , enter the digits in sequence that are to be added, and then depress the Add + key. The sum will appear on the display (X) register. Serial addition can also be performed by alternately entering numbers and depressing the Add + key. Example: $4 + 27 + 851 + 37 = 919$
−	Subtract	To subtract a number (subtrahend) from the value appearing on the display register (minuend), terminate the number string by depressing ENTER ↑ , enter the

HP-45 Keyboard **Summary of Most Used Functions**

Item	*Name*	*Description of Function*
		digits of the subtrahend in sequence, and then depress the Subtract [−] key. The remainder will appear on the display (X) register. Serial subtraction can also be performed by alternately entering subtrahends and depressing the Subtract [−] key. Example: 83.271 - 26.059 = 57.212
[×]	Multiply	To multiply a number (multiplier) by the value appearing on the display (X) register (multiplicand), terminate the number string by depressing [ENTER ↑], enter the digits of the multiplier in sequence, and then depress the Multiply [×] key. The product will appear on the display (X) register. Serial multiplication can also be performed by alternately entering multipliers and depressing the Multiply [×] key. Example: 0.9725 × 19.73 = 19.19
[÷]	Divide	To divide the value appearing on the display register (dividend) by a number (divisor), terminate the number string by depressing [ENTER ↑], enter the digits of the divisor in sequence, and then depress the Divide [÷] key. The quotient will appear on the display (X) register. Serial division can also be performed by alternately entering divisors and depressing the Divide [÷] key. Example: 8.072 ÷ 0.05971 = 135.2
[FIX]	Fixed Decimal Notation Display Mode	Depress the Fix [FIX] key and then the desired number key to specify the number of decimal places (0-9) to which the display is to be rounded. The display is left-justified and includes trailing zeros within the setting specified. Unless altered, the calculator is preset to display [FIX] 2 automatically.

HP-45 Keyboard		Summary of Most Used Functions
Item	*Name*	*Description of Function*
SCI	Scientific Notation Display Mode	Depressing the Gold Key ■ , the *alternate function* SCI key, and then the desired number key (0–9), results in the value being displayed in scientific notation. The display is left-justified and includes trailing zeros.
CHS	Change Sign	To change the sign of a number previously entered, depress the Change Sign CHS key. The number, preceded by a minus (-) sign, will appear on the display.
EEX	Enter Exponent	To indicate a power of ten, depress the Enter Exponent EEX key and then the numeric key denoting the desired power of ten.
1/x	Reciprocal	To obtain the reciprocal of a number in the display (X) register, depress the Reciprocal 1/x key. Example: $\frac{1}{15.62} = 6.402(10)^{-2}$
x^2	Square of x	To obtain the square of a number in the display (X) register depress the Square of x x^2 key. Example: $(0.00594)^2 = 3.53(10)^{-5}$
$\sqrt{x}$	Square Root of x	To obtain the square root of a number in the display (X) register, depress the Gold Key ■ and then the *alternate function* Square Root of x $\sqrt{x}$ key. The x in the symbol represents the currently displayed value. Example: $\sqrt{0.06057} = 0.2461$
ln	Natural Logarithm	To obtain the natural logarithm ($\log_e$) of a value appearing on the display (X) register, depress the Natural Logarithm ln key. Example: ln 1.026 = 0.0257

HP-45 Keyboard | **Summary of Most Used Functions**

Item	*Name*	*Description of Function*
e^x	Natural Antilogarithm	To obtain the natural antilogarithm (antilog$_e$) of a number appearing on the display (X) register, depress the Natural Antilogarithm e^x key. (To display the value of e, press 1 e^x.) Example: antilog$_e$ 0.0257 = 1.026 or $e^{0.0257}$ = 1.026
log	Common Logarithm	To obtain the common logarithm (log$_{10}$) of a number appearing on the display (X) register, depress the Gold Key ■, then the *alternate function* Common Logarithm log key. Example: log 11.91 = 1.0759 or: $10^{1.0759}$ = 11.91
10^x	Common Antilogarithm	To obtain the common antilogarithm of a number appearing on the display (X) register, depress the Gold Key ■, then the *alternate function* Common Logarithm 10^x key. Example: antilog 1.0759 = 11.91
DEG RAD GRD	Angular Modes: degree, radian, or decimal grads	Trigonometric functions can be performed in any one of three angular modes: decimal degrees, decimal radians, and decimal grads (a 100th part of a right angle in the centesimal system of measuring angles). To select a mode, depress the Gold Key ■, then the desired *alternate function* key DEG, RAD, or GRD. The mode selected will remain operative until a different mode is selected, or until the calculator is turned off; when turned back on, the calculator automatically defaults to the decimal degree mode.
→D.MS	Decimal Mode to Deg-Min-Sec Mode Conversion	To convert an angle appearing on the display register from decimal mode to degrees-minutes-seconds mode, depress the Gold Key ■, then the *alternate function* key →DMS. This feature is also

HP-45 Keyboard **Summary of Most Used Functions**

Item	*Name*	*Description of Function*
		useful in calculating problems dealing with time (hours-minutes-seconds). Example: 33.367° = 33°22′01″
D.MS→	Deg-Min-Sec to Decimal Mode Conversion	To convert an angle appearing on the display register from deg-min-sec to decimal mode, depress the Gold Key ■, then the *alternate function* key DMS→ . This feature is also useful in calculating problems dealing with time (hours-minutes-seconds). Example: 33°22′01″ = 33.367°
sin cos tan	Trigonometric Functions Sine Cosine Tangent	To obtain the sine, cosine, or tangent of an angle appearing on the display (X) register, depress the appropriate trigonometric function SIN , COS , or TAN key. Example: sin 26.5° = 0.4462
SIN^{-1} COS^{-1} TAN^{-1}	Inverse Trigonometric Functions Arc Sine Arc Cosine Arc Tangent	To obtain the arc sin, arc cos, or arc tan of a trigonometric function appearing on the display (X) register, depress the Gold key ■, and then the appropriate *alternate* trigonometric *function* key, SIN^{-1}, COS^{-1}, or TAN^{-1}. Example: arc sin 0.3392 = 19.8281° Example: arc cos 0.9065 = 24.9739° Example: arc tan 1.7256 = 59.9073°
y^x	Powers of Numbers	To raise a number appearing on the display (X) register to a real power, terminate the number string and move the value to the Y register by pressing ENTER ↑ , enter the desired power into the X register (using the numeric key set), depress the Gold Key ■, and then the *alternate function* Powers of Numbers y^x key. Example: $(0.232)^{0.0904} = 0.8763$

HP-45 Keyboard **Summary of Most Used Functions**

Item	*Name*	*Description of Function*
1/x y^x	Roots of Numbers	To extract the real root of a number appearing on the display (X) register, terminate the number string and move the value to the Y register by pressing ENTER ↑ , enter the desired root into the X register (using the numeric key), depress the reciprocal 1/x key, the Gold Key , and then the *alternate function* Roots of Numbers y^x key. Example: $\sqrt[0.75]{2.37} = 3.16$ or: $(2.37)^{1/0.75} = 3.16$
STO	Register Storage	To store a number appearing on the display (X) register (whether the result of a calculation or a keyboard entry), depress the Register Storage STO key, and then depress a number key (1 to 9) to specify the particular storage register which will receive the value.
RCL	Retrieve Storage	To retrieve a number which is stored in one of the nine storage registers, depress the Retrieve Storage RCL key, and then depress the applicable number key to specify the particular storage register from which the value will be taken.
R↓	Roll Down Stack	Depress the key R↑ each time you wish to roll down the stack one register.
T Z Y X	Operational Stack Registers: Top Register; 3rd Register; 2nd Register; Bottom Register (This is the register that is displayed)	The four temporary memory locations are arranged in the form of a "vertical stack." When a number is keyed into the calculator, it goes into the bottom register (X register) and is displayed. If ENTER ↑ is depressed, the displayed number is duplicated into the next higher (Y) register. Simultaneously, the value that was previously stored in the Y register is now moved up to the Z register and the value that was located in the Z register is moved up to the T register. (The contents that were previously located

HP-45 Keyboard		Summary of Most Used Functions
Item	*Name*	*Description of Function*
		in the T register are discarded (overwritten) so that the register can accept the incoming value.) When [+] is depressed, the value in register X (x) is added to the value in register Y (y), and the entire stack drops to display the answer in X. This same procedure occurs for [−], [×], and [÷]. Example: Load the stack by depressing 1 [ENTER ↑] 2 [ENTER ↑] 3 [ENTER ↑] 4. (The stack now contains $x = 4$, $y = 3$, $z = 2$, $t = 1$.) To review the contents of the stack depress [R↓] four times. The fourth [R↓] returns the stack to its original position ($x = 4$, $y = 3$, $z = 2$, and $t = 1$). Note: the stack is raised and t is lost when a keyboard entry or [RCL] operation follows [R↓], unless that entry follows [ENTER ↑], [CL x], or [Σ +].
[x⇄y]	Register Exchange	To interchange the values located in the X and Y registers, depress the Register Exchange [$x \lessgtr y$] key. It is often desirable to perform this exchange before [−], [÷], and [y^x] operations.
[LAST x]	Recall Last x Value	The last input value of a calculation is automatically stored in the "Last x" register when a function is executed. Recall of the value in this register is initiated by depressing the Gold Key ■ and then the *alternate function* Last x [LAST x] key. This feature provides a handy error correction device, as well as a facility for reusing the same value in multiple calculations. This register is cleared only when the calculator is turned off when a new value replaces (or overwrites) the previous one.

HP-45 Keyboard **Summary of Most Used Functions**

Item	*Name*	*Description of Function*
→P	Rectangular-to-Polar Conversion	To convert values in the X and Y registers (representing rectangular x and y coordinates, respectively) to polar r,θ coordinates (magnitude and angle, respectively), depress the Rectangular-to-Polar Conversion →P key to obtain r—and the Register Exchange $x \lessgtr y$ key to obtain θ.
→R	Polar-to-Rectangular Conversion	To convert values in the X and Y registers representing polar r,θ coordinates (magnitude and angle, respectively) to rectangular x and y coordinates, depress the Gold Key ■ and then the *alternate function* Polar-to-Rectangular Conversion →R key to obtain the x-coordinate—and the Register Exchange $x \lessgtr y$ key to obtain the y-coordinate.

A number of other specialized keys are provided on this calculator, but their use will not be discussed here. Working the problems suggested below will enable you to gain considerable skill and confidence.

General Examples of Calculator Capability

	Keyboard Entry	*Depress*	*Display Register*
Situation 1:			
12.32 - 7 + 1.6 = 6.92	12.32		12.32
		ENTER ↑	12.32
	7		7.
		−	5.32
	1.6		1.6
		+	6.92

Situation 2:

$-5.35 - (-4.2) - 3.1 = -4.25$

Keyboard Entry	Depress	Display Register
5.35	[CHS]	-5.35
	[ENTER ↑]	-5.35
4.2	[CHS] [−]	-1.15
3.1	[−]	-4.25

Situation 3:

$\left[\frac{(2+4)^5}{7} + 3\right] 6 = 43.714285$

Keyboard Entry	Depress	Display Register
2	[ENTER ↑]	2.00
4	[+]	6.00
5	[×]	30.00
7	[÷]	4.29
3	[+]	7.29
6	[×]	43.71

Situation 4:

$(8.7)^{2.6} + 34.7 = 311.87$

Keyboard Entry	Depress	Display Register
8.7	[ENTER ↑]	8.70
2.	■ [y^x]	277.17
34.7	[+]	311.87

Situation 5:

$\frac{1}{\frac{1}{560} + \frac{1}{390} + \frac{1}{670}} = 171.16$

Keyboard Entry	Depress	Display Register
	[FIX] [5]	0.00000000
560		560.

	Keyboard Entry	*Depress*	*Display Register*
Situation 5 (continued)			
		1/x	0.00179
	390		390.
		1/x	0.00256
		+	0.00435
	670		670.
		1/x	0.00149
		+	0.00584
		1/x	171.16388

	Keyboard Entry	*Depress*	*Display Register*
Situation 6:			
Convert the point (7,24) into polar coordinates.		■ DEG	
where Vector R = $\sqrt{x^2 + y^2}$	7	ENTER	7.00
σ = arc tan y/x	24	→P	25.00 (vector R)
and $x = 7$; $y = 24$		$x \lessgtr y$	16.26° (angle θ)

	Keyboard Entry	*Depress*	*Display Register*
Situation 7:			
A force of 8 newtons is acting at an angle of 120°. What are its rectangular components?		■ DEG	
	120	ENTER ↑	120.00
	8		8.
		■ →R	-4.00 (x-coordinate)
		$x \lessgtr y$	6.93 (y-coordinate)

Note: By combining the polar/rectangular function with the accumulation function, [Σ+] or [Σ−], you can add and subtract vector components. The sums of these are contained in storage register R_7 and R_8:

$$r_7 = x_1 \pm x_2 \pm \cdots \pm x_n = \Sigma x$$

$$r_8 = y_1 \pm y_2 \pm \cdots \pm y_n = \Sigma y$$

To display the contents of registers R_7 and R_8, press [RCL] [Σ+] to obtain the sum of x-coordinates (register 7); then press [x⪋y] to obtain the sum of y-coordinates (register 8).

	Keyboard Entry	*Depress*	*Display Register*
Situation 8:			
A traffic count at a street intersection indicated 8 Chryslers, 20 Fords, 17 Chevrolets, and 43 Volkswagens. What percentage of the total was represented by each manufacturer?	8	[ENTER ↑]	8.00
	20		20.
		[+]	28.00
	17		17.
		[+]	45.00
	43		43.
		[+]	88.00 (total)
		[STO] 1	88.00
	8	[RCL] 1 [÷]	0.09 (9% of total)
	20	[RCL] 1 [÷]	0.23 (23% of total)
	17	[RCL] 1 [÷]	0.19 (19% of total)
	43	[RCL] 1 [÷]	0.49 (49% of total)

Situation 9:

Sum two vectors having polar coordinates of 8 N at 30° and 12 N at 60°, respectively.

Keyboard Entry	*Depress*	*Display Register*
	■ CLEAR	
	■ DEG	
60	ENTER ↑	60.00
12		12.0
	■ →R	6.00
	Σ+	1.00
30	ENTER ↑	30.00
8		8.00
	■ →R	6.93
	Σ+	2.00
	RCL Σ+	12.93
	→P	19.35 N (magnitude)
	$x \lessgtr y$	48.07° (direction)

Practice Problems

Multiplication	pages 30 and 31
Division	pages 33 and 34
Multiplication and Division	pages 36 and 37
Squares	page 45
Square Roots	page 45
Cubes	page 48
Cube Roots	page 48
Sines	page 51
Cosines	pages 52 and 53
Tangents	page 55
Powers of Numbers	page 63
Roots of Numbers	page 65

THE ROCKWELL INTERNATIONAL UNICOM 202/SR, BOWMAR MX 100, LLOYD'S ELECTRONICS ACCUMATIC 99, AND SIMILAR CALCULATORS

Several manufacturers produce electronic hand calculators that utilize a Rockwell International trig function semiconductor chip in their circuit design. However, in each case there are some minor modifications that have been made that pertain more to marketing aspects and user convenience than to actual procedures of use or overall capability. In addition to Rockwell International's Unicom 202/SR, these include Lloyd's Electronics Accumatic 999, Bowmar's MX100, Sears' Electronic Slide Rule, Sharp's PC–1801, and similar calculators manufactured by Rapid Data, Keystone, and Summit. There are some differences in the action and life expectancy of the various keyboards associated with these calculators, and some care should be exercised in selecting the most desirable one. With only slight variations in operational procedure, however, these hand calculators may be operated using identical "address instructions." However, because of this similarity only the manipulative instructions for the Unicom 202/SR will be given here. In most cases the variations that have been made by the different manufacturers (usually these are nothing more than interchanged key locations) will be apparent to the user. These are all versatile calculation instruments that have been designed especially for the use of engineers, scientists, and technicians. They not only perform the four basic functions of arithmetic, but they have a number of other important features, including the capability to compute natural and common logarithms (used also to obtain the roots and powers of numbers) and trigonometric functions (sines, cosines, and tangents of angles and arc functions) and one fully addressable memory for storing data or accumulating results. These calculators have algebraic logic to allow us to compute the way that we have been taught to *think*. They can accept a numerical input of as many as eight digits. They also provide a "second function" key that allows each key to have two separate functions or uses (much like some typewriter keys). The first function of each key is printed on the surface of its key top, while the second function (referred to as the "alternate function") is usually imprinted on the keyboard immediately above each key. The operation of each key is described below.

Unicom 202/SR	**Keyboard**	**Summary of Most Used Functions**
Item	*Name*	*Description of Function*
ON / OFF	Power Switch	The switch should be turned off when the calculator is not in use. Be certain that the small switch is positioned down as far as possible (for "off" position), to avoid discharging the batteries while the calcu-

Unicom 202/SR	**Keyboard**	**Summary of Most Used Functions**
Item	*Name*	*Description of Function*
		lator is not in use. All internal storage of calculations within the machine are erased when the power switch is moved to the "off" position.
[0] to [9]	Numeric Key Set	Each digit of a number must be keyed in sequence, as desired, by depressing the appropriate number key. Digits entered in this manner are stored initially in the X register—the display register.
[•]	Decimal Point	The decimal point symbol must be depressed in the sequential position in which it occurs. Otherwise the calculator register will place a decimal to the right of the last value entered. For example, 314.32 would be keyed as [3] [1] [4] [•] [3] [2].
▬	Display Register	The display shows the contents of the entry (called the X register). When an answer exceeds 8 whole numbers, an overflow will occur and the display register will indicate the 8 most significant digits in the answer. Depress [C] and continue. The decimal point will appear 8 places to the left of the correct position. Multiply the answer in the display by 10^8 to obtain a decimally correct answer.
[C]	Clear	To erase the value shown on the display register (X register), depress the Clear [C] key. This will not affect the memory register. Depressing [C] after actuating a function key or after clearing an entry will reset any calculating mode and erase the entry register (X register). Depressing [C] during an overflow will reset the error condition. (An overflow occurs when the result of a calculation has exceeded the 8-digit capacity of the calculator, or if a mathematically impossible procedure has been tried.) The number in the display is correct if multiplied by 10^8, and may be used in further calculations. Chain and constant modes are not affected by overflowing.

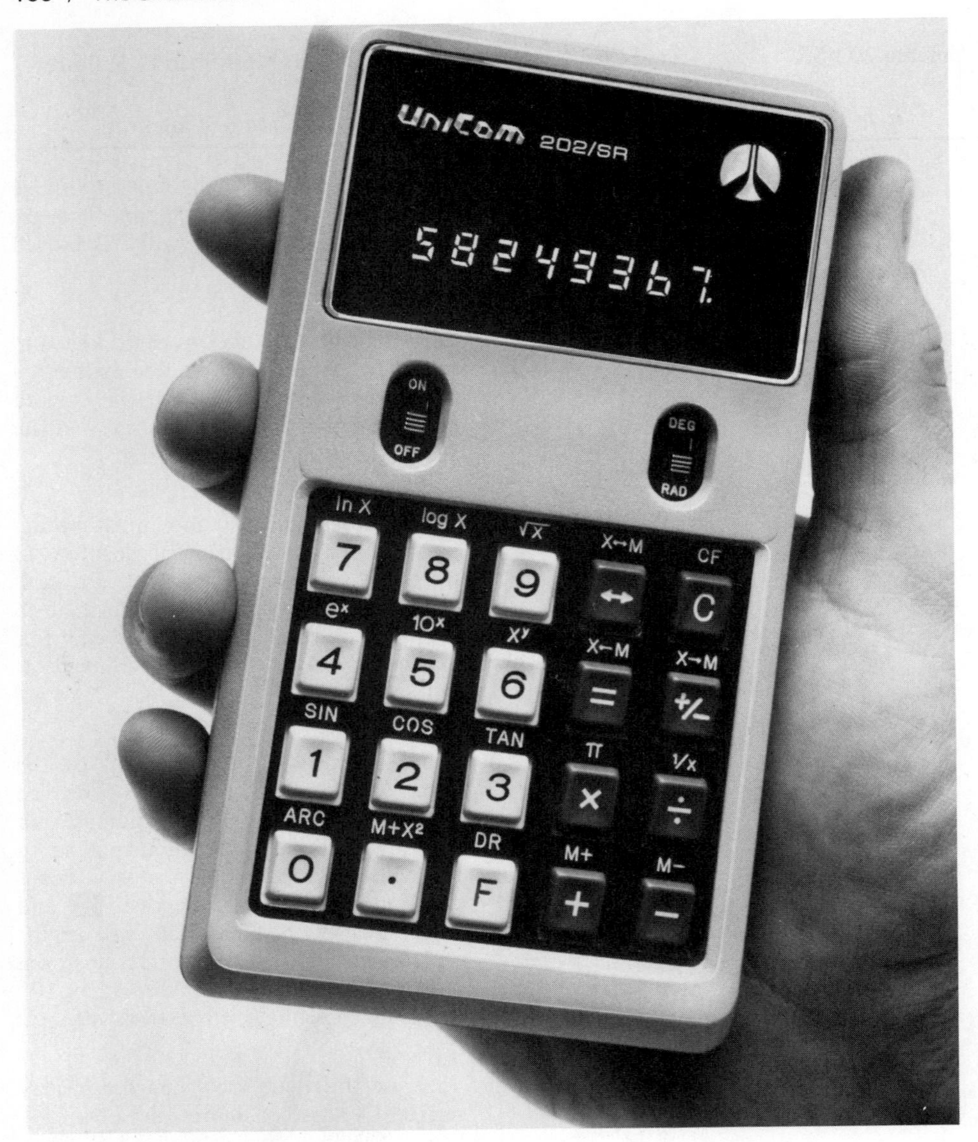

Illustration 3-2. The UNICOM 202/R Calculator. (Courtesy of UNICOM Systems, Inc.)

Unicom 202/SR	**Keyboard**	**Summary of Most Used Functions**
Item	*Name*	*Description of Function*
F	Function Key	The Function F key should be depressed prior to actuating alternate functions. This key is somewhat analogous to the "key shift" of a typewriter.

Illustration 3-3. The Bowmar MX100 Calculator. (Courtesy of Bowmar Consumer Products Division.)

Unicom 202/SR	Keyboard	Summary of Most Used Functions
Item	*Name*	*Description of Function*
[ ] (dashed box)	Alternate Function	*Alternate functions* are printed directly above their keyed locations. They are indicated like this, [F] [$\sqrt{x}$], in this text.
[CF]	Clear Alternate Function	To clear the *alternate function* mode that has been entered and restore the previous condition, depress [F] and then [CF].

Illustration 3–4. The Lloyd's Electronics Accumatic 999 Calculator. (Courtesy of Lloyd's Electronics, Inc.)

Unicom 202/SR *Item*	**Keyboard** *Name*	**Summary of Most Used Functions** *Description of Function*
DR	Digit Recall	To erase (recall) the last digit entered and shown on the display register, depress the Function F key and then the *alternate function* Digit Recall DR key. If more than one digit has been entered, depressing this combination of keys will eliminate the last digit and terminate the number entry mode. If a single digit has been entered, depressing this combination of keys will recall the previous result to the display.

Unicom 202/SR	**Keyboard**	**Summary of Most Used Functions**
Item	*Name*	*Description of Function*
		Example: If 12345 has been entered incorrectly as 12346, the mistake may be corrected by depressing in sequence [F], [DR], [CF], and [5]. The keys [F] and [DR] eliminate the 6, and [CF] eliminates or clears the *alternate function* mode and restores the number entry mode. Depressing the [5] key corrects the error.
[=]	Equals	To perform the previous arithmetic operation entered ([+], [−], [×], or [÷]) depress the Equals [=] key. This action will also store a constant in the working register and terminate the calculation.
[+]	Add	Depressing [+] executes the previously established condition (if any). To add a number to the value shown in the display register, depress the Add [+] key, enter the number, and then depress [+], [−], [×], [÷], or [=]. The sum will appear on the display register. Example: 215.88 + 49.27 = 265.15
[−]	Subtract	Depressing [−] executes the previously established condition (if any). To subtract a number (subtrahend) from the value appearing on the display register (minuend), depress the Subtract [−] key, enter the subtrahend, and then depress [+], [−], [×], [÷], or [=]. The remainder will appear on the display register. Example: 0.97211 - 0.02199 = 0.95012
[×]	Multiply	Depressing [×] executes the previously established condition (if any). To multiply a number (multiplier) by the value appearing on the display register (multiplicand), depress the Multiply [×] key, enter the multiplier, and then depress any

Unicom 202/SR	Keyboard	Summary of Most Used Functions
Item	*Name*	*Description of Function*
		of the keys [+], [−], [×], [÷], or [=]. The product will appear on the display register. Example: 10,925 × 12.85 = 140,386.25
[÷]	Divide	Depressing [÷] executes the previously established condition (if any). To divide the value appearing on the display register (dividend) by a number (divisor), depress the [÷] key, enter the divisor, and then depress any of the keys [+], [−], [×], [÷], or [=]. The quotient will appear on the display register. Example: 8.072 ÷ 0.05971 = 135.18673
[+/-]	Change Sign	To change the sign of a number previously entered and shown on the display register, depress the Change Sign [+/-] key. The presence of a negative quantity is noted by the appearance of a lighted signal in the "Minus Displày," which is located at the upper right corner, above the display register.
[1/x]	Reciprocal	To obtain the reciprocal of a number in the display register, depress [F] and then [1/x]. Example: $\frac{1}{15.62} = 0.0640204$ 15.62 [F] [1/x]
[√x]	Square Root of x	To obtain the square root of a number in the display register, depress [F] and then [√x]. Example: $\sqrt{0.06057} = 0.2461097$ .06057 [F] [√x]
[ln x]	Natural Logarithm	To obtain the natural logarithm ($\log_e$) of a value appearing on the display register depress [F] and then [ln x]. Example: 1n 1.026 = 0.025668 1.026 [F] [ln x]

Unicom 202/SR Item	Keyboard Name	Summary of Most Used Functions — Description of Function
e^x	Natural Antilogarithm	To obtain the natural antilogarithm (antilog$_e$) of a number appearing on the display (X) register, depress [F] and then [e^x]. Example: antilog$_e$ 0.0257 = 1.026033 or $\epsilon^{1.026033} = 0.0257$ .0257 [F] [e^x]
logX	Common Logarithm	To obtain the common logarithm ($\log_{10}$) of a number appearing on the display register, depress [F] and then [log x]. Example: log 11.91 = 1.075912 11.91 [F] [log x]
10^x	Common Antilogarithm	To obtain the common antilogarithm of a number appearing on the display (X) register, depress [F] and then [10^x]. Example: antilog 1.0759 = 11.90968 or $(10)^{1.0759}$ = 11.90968 1.0759 [F] [10^x]
DEG / RAD	Angular Mode	Prior to entering an angle into the display register, select the angular mode (degrees or radians) to be used by appropriately positioning the angular mode switch.
SIN COS TAN	Trigonometric Functions: Sine, Cosine, Tangent	To obtain the sine, cosine, or tangent of an angle appearing on the display register, depress the Function [F] key and then the appropriate *alternate Trigonometric Function* [SIN], [COS], or [TAN]. Example: sin 26.5° = 0.446198 [DEG] 26.5 [F] [SIN] cos 15.9° = 0.961471 [DEG] 15.9 [F] [COS] tan 67.5° = 2.414214 [DEG] 67.5 [F] [TAN]

Unicom 202/SR Keyboard Summary of Most Used Functions

Item	*Name*	*Description of Function*
ARC	Inverse Trigonometric Functions	To obtain the arc sin, arc cos, or arc tan of a trigonometric function appearing on the display register, depress in sequence the Function F key, the Arc ARC key, and then the appropriate *alternate trigonometric function* SIN, COS, or TAN.
SIN^{-1}	Arc Sine	Example: arc sin 0.3392 = 19.82814°
COS^{-1}	Arc Cosine	DEG .3392 F ARC SIN
TAN^{-1}	Arc Tangent	arc cos 0.9065 = 24.97392°
		DEG .9065 F ARC COS
		arc tan 1.7256 = 59.90733°
		DEG 1.7256 F ARC TAN
x^Y	Powers of Numbers	To raise a number appearing on the display (X) register to a real power (y), depress F and then x^y, enter the power y, and depress =.
		Example: $0.232^{0.0904} = 0.876275$
		.232 F x^y .0904 =
1/X x^Y	Roots of Numbers	To extract the real root of a number appearing on the display (X) register, depress F and then x^y, enter the root y and depress F 1/x =.
		Example: $\sqrt[4]{279{,}841} = 23$
		279841 F x^y 4 F 1/x =
↔	Exchange Registers	To exchange the number in the display (X) register with the number in the working (Y) register, depress the Exchange Register ↔ key.
M+	Add to Memory	To add a number appearing on the display (X) register to the number in the memory, depress F and then M+. The memory will now contain the sum of the previously stored value and the number that was transferred from the display (X) register.

Unicom 202/SR **Keyboard** **Summary of Most Used Functions**

Item	*Name*	*Description of Function*
M−	Subtract from Memory	To subtract a number appearing on the display (X) register from the number in the memory, depress F and then M− .
X←M	Memory Display	To display the number currently stored in the memory, depress F and then X←M . The contents of the memory are not altered by this operation. However, this action does clear the display (X) register before performing the transfer from the memory.
X→M	Display to Memory (also Memory Clear)	To replace the number in the memory with the number appearing on the display (X) register, depress F and then X→M . This action clears the memory to zero before performing the transfer from the display register. However, it does not affect the value previously stored in the display (X) register.
X↔M	Exchange with Memory	To exchange the number appearing on the display (X) register with the number in the memory, depress F and then X↔M .
$M+X^2$	Square to Memory	To add the square of the number appearing on the display (X) register to the memory, depress F and then $M+x^2$. The display (X) register is not altered by this operation. Also, execution of the Square to Memory key does not clear the memory.

General Examples of Calculator Capability

	Keyboard Entry	*Depress*	*Display Register*
Situation 1:		C	0.
12.32 − 7 + 1.6 = 6.92	12.32		12.32
		−	12.32
	7		7.
		+	5.32
	1.6		1.6
		=	6.92

Situation 2:

$-5.35 - (-4.2) - 3.1 = -4.25$

Keyboard Entry	Depress	Display Register
	C	0.
5.35		5.35
	+/-	-5.35
	−	-5.35
4.2		4.2
	+/-	-4.2
	−	-1.15
3.1		3.1
	=	-4.25

Situation 3:

$\left[\frac{(2+4)5}{7} + 3\right] 6 = 43.714285$

Keyboard Entry	Depress	Display Register
	C	0.
2	+	2.
4	×	6.
5	÷	30.
7	+	4.2857142
3	×	7.2857142
6	=	43.714285

Situation 4:

$(8.7)^{2.6} + 34.7$

Keyboard Entry	Depress	Display Register
	C	0.
8.7		8.7
	F x^y	2.163323

Situation 4 (cont)

	Keyboard Entry	*Depress*	*Display Register*
	2.6		2.6
		=	277.1724
		+	277.1724
	34.7		34.7
		=	311.8724

Situation 5:

	Keyboard Entry	*Depress*	*Display Register*
$34.7 + (8.7)^{2.6}$		C	0.
	34.7		34.7
		F X→M	34.7
	8.7		8.7
		F x^y	2.163323
	2.6		2.6
		=	277.1724
		F M+	277.1724
		F X←M	311.8724

Situation 6:

	Keyboard Entry	*Depress*	*Display Register*
$\dfrac{1}{\dfrac{1}{560}+\dfrac{1}{390}+\dfrac{1}{670}} = 171.16388$		C	0.
	560		560
		F 1/x +	0.0017857

Situation 6 (cont)

Keyboard Entry	Depress	Display Register
390		390
	F 1/x +	0.0043498
670		670
	F 1/x =	0.0058423
	F 1/x	171.16546

Situation 7:

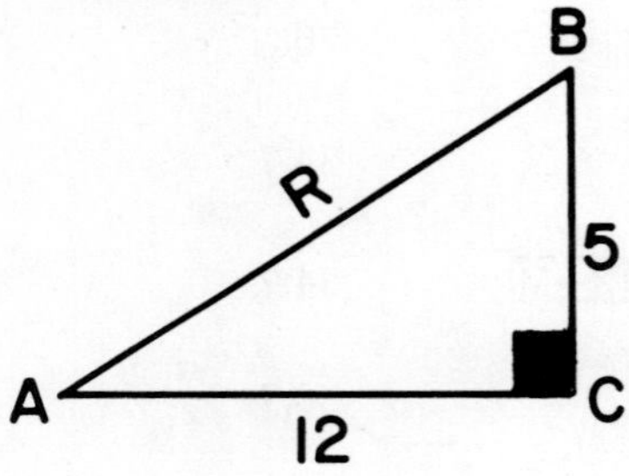

B, R, 5, A, C, 12

$R^2 = (5)^2 + (12)^2$

$R = \sqrt{169} = 13$

Keyboard Entry	Depress	Display Register
	C	0.
	F X→M	0.
5		5.
	F $M+x^2$	5.
12		12.
	F $M+x^2$	12.
	F X←M	169.
	F $\sqrt{x}$	13.

Situation 8:

Convert the point (7,24) into polar coordinates.

where: Vector $V = \sqrt{x^2 + y^2}$

$\sigma = \text{arc tan} \frac{y}{x}$

and: $x = 7$; $y = 24$

Keyboard Entry	Depress	Display Register
	C	0.
	F X→M	0.
7		7.
	F $M+x^2$ ÷	7.

Situation 8:

Keyboard Entry	*Depress*	*Display Register*
24		24.
	[F] [M+x^2]	24.
	[=]	0.2916666
	[F] [ARC] [TAN]	16.2602° (angle σ)
	[F] [X←M]	625
	[F] [$\sqrt{x}$]	25 (vector V)

Situation 9:

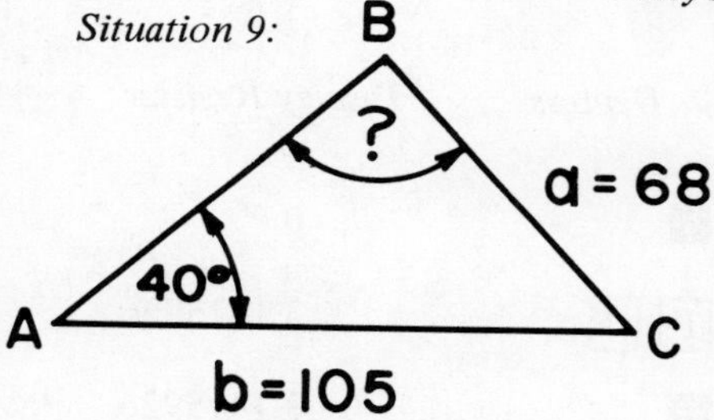

B
?
a = 68
40°
A
C
b = 105

Solve for angle B.

$$\frac{a}{\text{Sin A}} = \frac{b}{\text{Sin B}} = \frac{c}{\text{Sin C}}$$

$$\text{Sin B} = \frac{\text{b Sin A}}{a}$$

$$\text{B} = \text{Arc Sin}\left[\frac{\text{b Sin A}}{a}\right]$$

$$\text{B} = \text{Arc Sin}\left[\frac{105 \times \text{Sin } 40°}{68}\right]$$

Keyboard Entry	*Depress*	*Display Register*
	[C]	0. 0.
	[DEG/RAD]	0.
40.		40.
	[F] [SIN]	0.642788
	[×]	0.642788
105		105.
	[÷]	67.49274
68		68.
	[=]	0.9925402
	[F] [ARC] [SIN]	82.99719° (angle *B*)

Situation 10:

Solve for the hyperbolic sine of 1.1 (sinh 1.1).

where:

$$\sinh a = \frac{e^a - e^{-a}}{2}$$

Keyboard Entry	*Depress*	*Display Register*
	C	0.
1.1		1.1
	F e^x	3.004165
	−	3.004165
	F 1/x	0.3328711
	÷	2.6712938
2.		2.
	=	1.3356469 (sinh 1.1)

Situation 11:

Solve for the hyperbolic tangent of $\pi/3$ (tanh $\pi/3$).

where

$$\tanh a = \frac{e^a - e^{-a}}{e^a + e^{-a}}$$

Keyboard Entry	*Depress*	*Display Register*
	C	0.
	F π	3.1415926
	÷	3.1415926
3		3.
	=	1.0471975
	F e^x	2.849652
	F X→M	2.849652
	−	2.849652
	F 1/x	0.35092
	F M+	0.35092
	÷	2.498732
	F X←M	3.200572
	=	0.7807141 (tanh $\frac{\pi}{3}$)

Situation 12:	*Keyboard Entry*	*Depress*	*Display Register*
Determine V_c		C	0.
if: r = 50 kilo ohms	.015		0.015
c = 0.1 microfarads		÷	0.015
t = 0.015 sec.	.000 000 1		0.0000001
V_1 = 25 volts		÷	15,000.
where:	50,000		50,000.
$V_c = V_1(1 - e^{-\frac{t}{rc}})$		=	3.
		+/-	-3.
		F [e^x]	0.049787
		−	0.049787
	1.		1.
		↔	0.049787
		×	0.950213
	25		25.
		=	23.755325 (V_c)

Situation 13:	*Keyboard Entry*	*Depress*	*Display Register*
How much will an investment of $1000 be worth in 8 quarters		C	0.
if interest is 7.1% annually and the interest is compounded	.071		0.071
quarterly?		÷	0.071
if: i = interest rate for a given interest period	4		4.
n = number of interest periods		+	0.01775

Situation 13 (cont)

where:

Future value = (Present Value)$(1 + i)^n$

Future value = $\$1000 \left[1 + \frac{0.071}{4}\right]^{8}$

Keyboard Entry	Depress	Display Register
1		1.
	=	1.01775
	F xy	0.017594
8		8.
	=	1.151139
	×	1.151139
1000		1000.
	=	$1151.139 (Future value)

Situation 14:

If the quarterly interest rate is 1.5%, how much money must be invested now to grow to $25,000 in 3 years (12 quarters)?

if: i = interest rate for a given interest period

n = number of interest periods

where:

$$\text{Present value} = \frac{\text{Future value}}{(1+i)^n}$$

$$\text{Present value} = \frac{\$25{,}000}{(1+0.015)^{12}}$$

Keyboard Entry	Depress	Display Register
	C	0.
1.015		1.015
	F xy	0.014888
12		12.
	=	1.195609
	÷	1.195609
25,000		25,000.
	÷	1.195609
	=	$20,909.845 (Present value)

Situation 15:

A loan of $3000 is to be repaid in 24 equal monthly installments. The annual interest rate is 7.5%. How much is each payment?

if: i = interest rate for a given interest period

n = number of interest periods

where:

Uniform payment =

$$\left[\frac{i(1+i)^n}{(1+i)^n - 1}\right] \text{[Principal Sum]}$$

Uniform payment =

$$\left[\frac{\frac{0.075}{12}\left\{1+\frac{0.075}{12}\right\}^{24}}{\left\{1+\frac{0.075}{12}\right\}^{24} - 1}\right] \$3000$$

Keyboard Entry	Depress	Display Register
	C	0.
.075		0.075
	÷	0.075
12		12.
	+	0.00625
	F X→M	0.00625
1		1.
	=	1.00625
	F x^y	0.00623
24		24.
	=	1.161277
	×	1.161277
	F X↔M	0.00625
	÷	0.0072579
1		1.
	F M-	1.
	F X←M	0.161277
	×	0.0450026
3000		3000.
	=	135.0078 (monthly loan payment)

Practice Problems

Multiplication	pages 30 and 31
Division	pages 33 and 34
Multiplication and Division	pages 36 and 37
Squares	page 45
Square Roots	page 45
Cubes	page 48
Cube Roots	page 48
Sines	page 51
Cosines	pages 52 and 53
Tangents	page 55
Powers of Numbers	page 63
Roots of Numbers	page 65

THE TEXAS INSTRUMENTS SR–50

One of the most recent entries into the electronic hand calculator field is the Texas Instruments' SR–50. This small calculator is capable of processing a wide range of problems from simple arithmetic to complex engineering and scientific calculations. As with the "Rockwell type" calculators (see pages 98 to 115) it was designed to use algebraic notation for data entry, rather than the Lukasciewicz or so-called "reverse Polish" notation, which the Hewlett-Packard calculators employ. Because algebraic logic was used in the design of the MOS solid state circuitry, numbers and algebraic functions are entered into the calculator in the same order that they are written on paper, and intermediate results may be obtained by depressing the "Equals" **=** key. One of this calculator's most desirable features is its ability to accept entries in scientific notation, which is not usually possible with many of the hand calculators that use algebraic notation. Also, the SR–50 retains 13 significant digits of accuracy for internal processing. These are rounded off to 10 digits for the display.

This calculator uses three registers for processing. The X register is the display register, and its contents are shown on the display. The Y register is the working register. It always stores those quantities awaiting completion of any processing (except intermediate functions). The Z register is the cumulative register for addition and subtraction. A memory (M register) is used for storage and for summing to memory. It is never used for processing. Because of this feature stored data does not get displaced during the processing of functions. Also, with the three process registers any sequence or combination of functions can be processed without overflowing the registers or losing data.

The general operation of the calculator is also simplified because the keyboard design does not require the use of an *alternate function* set of keys, as is the case with a number of other calculators. The single memory is adequate for general purpose use. The operations of the keys used most often are described below.

Illustration 3-5. The Texas Instruments Incorporated SR 50 Calculator. (Courtesy of Texas Instruments Incorporated.)

Item	*Name*	*Description of Function*
ON (switch)	Power Switch	The switch should be turned off when the calculator is not in use. All internal storage of calculations within the machine is erased when the power switch is moved to the "OFF" position. The power-on condition is indicated by the presence of a number in the display register. If, after turning the calculator on, a number appears on the display (before performing the first calculation), depress the Clear [C] key to assure register clearance.
[0] to [9]	Numeric Key Set	Each digit of a number must be keyed in sequence, as desired, by depressing the appropriate number key. Digits entered in this manner are stored initially in the X register–the display register.

Item	*Name*	*Description of Function*
[π]	Pi	To enter the constant pi (3.141592654) depress the Pi [π] key.
[•]	Decimal Point	The decimal point symbol must be depressed in the sequential position in which it occurs. Otherwise the calculator register will place a decimal to the right of the last value entered. For example, 314.32 could be keyed as [3] [1] [4] [·] [3] [2] .
[EE]	Enter Exponent	To enter a number into the calculator in scientific notation (as a number multiplied by 10 raised to some power), depress the appropriate numeric keys to enter the number, depress the Enter Exponent [EE] key, and then the numeric keys denoting the desired exponent of 10. The last two digits on the right side of the display are used to indicate exponents. After the Enter Exponent [EE] key has been depressed, the calculator will display all further results in scientific notation until the Clear [C] key is depressed.
▬	Display Register	The display shows the contents of the X register. In addition to power-on indication and numerical information, the display also provides indication of a negative number, decimal point, overflow, underflow, and error.
[CE]	Clear Entry	To erase an erroneous entry shown on the display register (X register) depress the Clear Entry [CE] key. This will not affect the working register (Y register), the cumulative register (Z register), or the memory.
[C]	Clear	To erase the value shown on the display register (X register), the value stored in the working register (Y register), and the value stored in the cumulative register (Z register), depress the Clear [C] key. This will not affect the memory register. (To clear the memory, a zero must be entered in place of the stored data.)

Item	*Name*	*Description of Function*
[=]	Equals	To obtain intermediate or final results of a computation, depress the Equals [=] key. This action instructs the calculator to complete the processing of all the previously entered data and algebraic functions.
[+]	Add	Depressing [+] executes the previously established condition (if any). To add a number to the value shown in the display register, depress the Add [+] key, enter the digits in sequence, and then depress any of the keys [+] , [−] , or [=] . The sum will appear on the display register. Example: 215.88 + 49.27 = 265.15
[−]	Subtract	Depressing [−] executes the previously established condition (if any). To subtract a number (subtrahend) from the value appearing on the display register (minuend), depress the Subtract [−] key, enter the digits of the subtrahend in sequence, and then depress any of the keys [+] , [−] , or [=] . The remainder will appear on the display register. Example: 0.97211 - 0.02199 = 0.95012
[×]	Multiply	Depressing [×] executes the previously established condition (if any). To multiply a number (multiplier) by the value appearing on the display register (multiplicand), depress the Multiply [×] key, enter the digits of the multiplier in sequence, and then depress any of the keys [+] , [−] , [×] , [÷] , or [=] . The product will appear on the display register. Example: 10,925 × 12.85 = 140,386.25
[÷]	Divide	Depressing [÷] executes the previously established condition (if any). To divide the value appearing on the display register (dividend) by a number (divisor), depress the Divide [÷] key, enter the digits of the

Item	*Name*	*Description of Function*
		divisor in sequence, and then depress any of the keys [+], [−], [×], [÷], or [÷]. The quotient will appear on the display register. Example: 8.072 ÷ 0.05971 = 135.18673
[+/−]	Change Sign	To change the sign of a number previously entered and appearing on the display register, depress the Change Sign [+/−] key. To enter a negative number, first enter the number and then press the Change Sign [+/−] key. To change the sign of the exponent appearing in the display, first depress the Enter Exponent [EE] key, and then the Change Sign [+/−] key.
[1/x]	Reciprocal	To obtain the reciprocal of a number appearing on the display register, depress the Reciprocal [1/x] key. Example: $\frac{1}{15.62} = 6.402(10)^{-2}$
[x^2]	Square of x	To obtain the square of a number in the display register, depress the Square of x [x^2] key. Example: $(4.2)^2 = 17.64$
[$\sqrt{x}$]	Square Root of x	To obtain the square root of a number in the display register, depress the Square Root of x [$\sqrt{x}$] key. Example: $\sqrt{0.06057} = 0.246109732$
[lnx]	Natural Logarithm	To obtain the natural logarithm ($\log_e$) of a value appearing on the display register, depress the Natural Logarithm [lnx] key. Example: ln 1.026 = 0.025667747
[e^x]	Natural Antilogarithm (*e* to the *x* power)	To obtain the natural antilogarithm (antilog_e) of a number appearing on the display register, depress the Natural Antilogarithm [e^x] key. Example: $\text{antilog}_e\ 0.0257 = 1.026033093$ or $_{\epsilon}1.026033092 = 0.0257$

Item	*Name*	*Description of Function*
log	Common Logarithm	To obtain the common logarithm ($\log_{10}$) of a number appearing on the display register, depress the Common Logarithm log key. Example: log 11.91 = 1.075911761
R D (switch)	Angular Mode	Select the angular mode (degrees or radians) desired by appropriately positioning the Angular Mode switch. The calculator interprets a displayed angle as being in either degrees or radians, depending upon the angular mode setting.
D/R	Degrees/Radians	To convert a displayed angle from degrees to radians (or vice versa) depress the Degrees/Radians D/R key.
sin cos tan	Trigonometric Functions	To obtain the sine, cosine, or tangent of an angle appearing on the display register, depress the appropriate Trigonometric Function sin , cos , or tan key. Example: sin 26.5° = 0.446197813 cos 15.9° = 0.961741309 tan 67.5° = 2.414213561
arc	Inverse Trigonometric Functions	To obtain the arc sin, arc cos, or arc tan of a trigonometric function appearing on the display register, depress the Inverse arc key and then the appropriate Trigonometric Function sin , cos , or tan key. Example: arc sin 0.3392 = 19.82814125° arc cos 0.9065 = 24.97392830° arc tan 1.7256 = 59.90734021°
hyp	Hyperbolic Functions	To obtain the hyperbolic sine, hyperbolic cosine, or hyperbolic tangent of an angle appearing on the display register, depress the Hyperbolic hyp key and then

Item	*Name*	*Description of Function*
		the appropriate Trigonometric Function sin , cos , or tan key. Example: sinh 1.1 = 1.3356469 cosh 0.8 = 1.3374385 $\tanh\frac{\pi}{3} = 0.7807141$
y^x	Powers of Numbers	To raise a number appearing on the display register to a real power, depress the Powers of Numbers y^x key, the numeric keys corresponding to the digits denoting the power, and then the Equals = key. (Depressing a function key rather than the Equals = key will produce the same effect.) Example: $(0.232)^{0.0904} = 0.876274379$
$\sqrt[x]{y}$	Roots of Numbers	To extract the real root of a number appearing on the display register, depress the Roots of Numbers $\sqrt[x]{y}$ key, the numeric keys corresponding to the digits denoting the root, and then the Equals = key. (Depressing a function key rather than the Equals = key will produce the same effect.) Example: $\sqrt[4]{279,841} = 23$
x:y	Exchange Registers	To exchange the number in the display (X) register with the number in the working (Y) register, depress the Exchange Register x:y key.
STO	Memory Storage	To replace the number in the memory with the number appearing on the display register, depress the Memory Storage STO key.
Σ	Sum to Memory	To add a number appearing on the display register to the number in the memory, depress the Sum to Memory Σ key. The use of this key does not affect the displayed quantity nor the previously processed data.

Item	*Name*	*Description of Function*
RCL	Memory Display	To display (recall) the number currently stored in the memory, depress the Memory Display RCL key. The contents of the memory is not altered by this operation. However, this action does clear the display register before performing the transfer from the memory.

General Examples of Calculator Capability:

Situation 1:

	Keyboard Entry	*Depress*	*Display Register*
12.32 - 7 + 1.6 = 6.92	12.32	⊟	12.32
	7	⊞	5.32
	1.6	⊜	6.92

Situation 2:

	Keyboard Entry	*Depress*	*Display Register*
-5.35 - (-4.2) - 3.1 = -4.25	5.35	+/- ⊟	-5.35
	4.2	+/- ⊟	-1.15
	3.1	⊜	-4.25

Situation 3:

	Keyboard Entry	*Depress*	*Display Register*
$\left[\frac{(2+4)\,5}{7} + 3\right] 6 =$	2	⊞	2.
	4	⊜ ⊠	6.
= 43.71428572	5	÷	30.
	7	⊞	4.285714286
	3	⊜ ⊠	7.285714286
	6	⊜	43.71428571

Situation 4:

$(8.7)^{2.6} + 34.7 = 311.8724475$

Keyboard Entry	Depress	Display Register
8.7	y^x	8.7
2.6	+	277.1724475
34.7	=	311.8724475

Situation 5:

$34.7 + (8.7)^{2.6} = 311.8724475$

Keyboard Entry	Depress	Display Register
34.7	+	34.7
8.7	y^x	8.7
2.6	=	311.8724475

Situation 6:

$$\frac{1}{\frac{1}{560}+\frac{1}{390}+\frac{1}{670}} =$$

$= 171.1638788$

Keyboard Entry	Depress	Display Register
560	1/x +	0.0017857142
390	1/x +	0.0043498168
670	1/x =	0.005842354
	1/x	171.1638788

Situation 7:

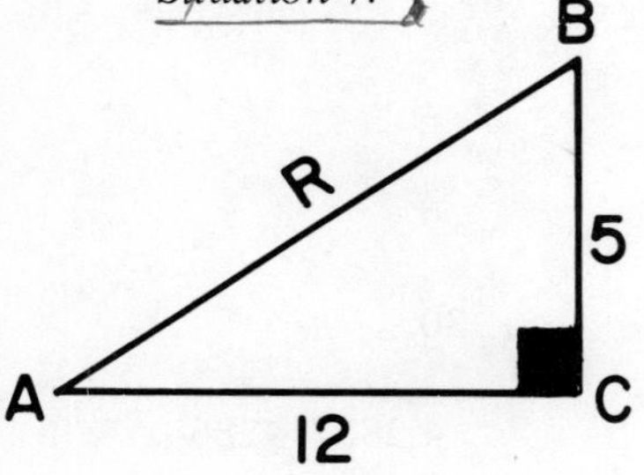

B
R
5
A
C
12

$R^2 = (5)^2 + (12)^2$

$R = \sqrt{169} = 13$

Keyboard Entry	Depress	Display Register
5	x^2 +	25.
12	x^2 =	169.
	$\sqrt{x}$	13.

Situation 8:

Convert the point (7,24) into polar coordinates

where: vector v = $\sqrt{x^2 + y^2}$

$\sigma = \text{arc tan } y/x$

and $x = 7$; $y = 24$

Keyboard Entry	Depress	Display Register
7	STO ÷	7.
24	= arc tan	16.26020471° (angle σ)
24	x² +	576.
	RCL x² =	625.
	√x	25. (vector v)

Situation 9:

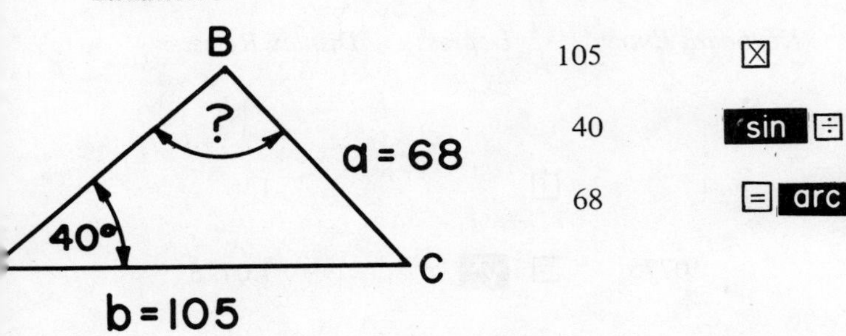

Solve for angle B.

$$\frac{a}{\sin A} = \frac{b}{\sin B} = \frac{c}{\sin C}$$

$$\sin B = \frac{b \sin A}{a}$$

$$B = \text{arc sin}\left[\frac{b \sin A}{a}\right]$$

$$B = \text{arc sin}\left[\frac{105 \times \sin 40^\circ}{68}\right]$$

Keyboard Entry	Depress	Display Register
105	×	105.
40	sin ÷	67.49269902
68	= arc sin	82.99696344 (angle B)

Situation 10:

Determine V_c

if $r = 3300$ ohms

$c = 47$ microfarads

$t = 250 \times 10^{-3}$ sec

$V_1 = 18$ volts

where:

$$V_c = V_1 \left[1 - e^{-\frac{t}{rc}}\right]$$

$$V_c = 18\left[1 - e^{-\frac{250 \times 10^{-3}}{3300 \times 47 \times 10^{-6}}}\right]$$

$V_c = 14.40872087$ volts

Keyboard Entry	Depress	Display Register
250	+/- EE	-250 00
3	+/- ÷	-2.5 -01
3300	÷	-7.575757576 -05
47	EE +/-	47 -00
6	= e^x +/- +	-1.995155075 -01
1	= ×	8.004844925 -01
	=	1.440872087 01 (V_c)

Situation 11:

If $15,000 is invested at 7¾% interest compounded annually, what will be the accumulated amount at the end of 8 years?

if: i = interest rate for a given interest period

n = number of interest periods

where:

Future Value = (Present Value) $(1 + i)^n$

Future Value = $15,000 $(1 + 0.0775)^8$

Future Value = 27,253.95

Keyboard Entry	Depress	Display Register
1	+	1.
.0775	= y^x	1.0775
8	×	1.816930146
15000	=	27253.95219

Situation 12:

What is the present value of the future amount \$35,570 in 13 years? The interest rate is 6.3% compounded quarterly.

if: i = interest rate for a given interest period

n = number of interest periods

where:

$$\text{Present Value} = \frac{\text{Future Value}}{(1+i)^n}$$

$$\text{Present Value} = \frac{\$35{,}570}{(1+\frac{0.063}{4})^{13\times 4}}$$

Present Value = \$15,782.24

Keyboard Entry	Depress	Display Register
4	×	4.
13	= STO	52.
.063	÷	0.063
4	+	0.01575
1	= y^x RCL = $^1/_x$ ×	.4436952589
35570	=	15782.24036

Situation 13:

Solve for h:

$$h = 0.023\left(\frac{0.423}{0.08}\right)(20{,}000)^{0.8} \times \left[\frac{0.76 \times 10.16}{0.423}\right]^{0.4}$$

h = 1072.380692

Keyboard Entry	Depress	Display Register
.76	×	.76
10.16	÷	7.7216
.423	y^x	18.25437352
.4	= STO	3.195558505
20000	y^x	20000.
.8	×	2759.459323
.423	×	1167.251294
.023	÷	26.84677975
.08	× RCL =	1072.380692

Situation 14:

Solve for *S*:

$$S = \frac{16}{\pi(1.3)^3} \times \left[875 + \sqrt{(875)^2 + (1500)^2}\right]$$

S = 6053.95673

Keyboard Entry	Depress	Display Register
875	x² +	765625
1500	x² =	
	√x +	1736.555499
875	= ×	2611.555499
16	÷	41784.88798
π	= STO	13300.54294
1.3	yˣ	1.3
3	÷ RCL x:y =	6053.95673

Practice Problems

Multiplication pages 30 and 31
Division pages 33 and 34
Multiplication and Division pages 36 and 37
Squares page 45
Square Roots page 45
Cubes page 48
Cube Roots page 48
Sines page 51
Cosines pages 52 and 53
Tangents page 55
Powers of Numbers page 63
Roots of Numbers page 65

4

the metric [SI] and other unit systems

Man interprets the universe in which he lives by evaluating those things that he perceives. Through experience he has learned that there are certain physical quantities that are unique and *fundamental* and that can be used to describe all other physical relationships. Among the fundamental dimensions most commonly recognized are *length, force,* and *time,* which are used extensively by peoples of all cultures, economic classes, and educational levels. Engineering and scientific calculations make use of measurements of all types and, therefore, use not only these, but other fundamental dimensions as well. Fundamental dimensions may be combined in numerous ways to form *derived dimensions;* it is by this means that man is able to portray accurately the physical laws of nature that he observes.

Some measurements are made with precise instruments, while others are the result of crude approximations. Regardless of the accuracy of the measurements or of the particular type of measuring instrument used, the measurements are themselves merely representative of certain comparisons previously agreed upon.

The length of a metal cylinder, for example, can be determined by laying it alongside a calibrated scale or ruler. The 12-in. ruler is known to represent one third of a yard, and a yard is recognized as being equivalent to 36.00/39.37 metre[1] –which used to be the distance between two marks on a

[1] The SI spelling "metre" is used instead of the more traditional spelling, "meter," used in the past in this country. Both spellings are acceptable.

platinum-iridium bar kept in a vault in Sèvres, France, but is now defined in terms of the wavelength of a particularly uniform monochromatic light. All these methods of measurements are comparisons. Other similar standards exist for the measurement of temperature, time, and force.

Physical quantities to be measured may be of two types: those concerned with *fundamental dimensions* of length (L), time (T), force (F), mass (M), electrical charge (Q), luminous intensity (I), and temperature (θ); and those concerned with *derived dimensions,* such as area, volume, pressure, or density. *Fundamental dimensions* may be subdivided into various sized parts, called *units.* The dimension *time* (T), for example, can be expressed in the units of seconds, hours, days, and so forth, depending upon the application to be made or the magnitude of the measurement. *Derived dimensions* are categorical descriptions of some specific physical characteristic or quality of an entity, and they are brought into being by combining *fundamental dimensions.* Area, therefore, is expressed dimensionally as length times length, or length squared (L^2), pressure as force per unit area (F/L^2, or FL^{-2}), and acceleration as length per time squared (L/T^2, or LT^{-2}).

Most measured quantities must be expressed in both magnitude and units. To state that an area was 146 would have no meaning. For example, an area could be tabulated as 146 mi^2 or 146 cm^2; a pressure could be recorded as 0.0015 dyne/cm^2 or 0.0015 lb$_f$/in.^{2}; an acceleration could be indicated as 159 in./sec^2 or 159 ft/sec^2, and so forth. However, some values used in engineering computations are dimensionless (without dimensions). These should be ignored in the unit balancing of an equation. *Radians,* π, *coefficient of friction, ratios,* and *per cent error* are examples of dimensionless quantities.

Equations involving measured quantities must be balanced dimensionally as well as numerically. Both dimensions and units can be multiplied and divided or raised to powers just like ordinary algebraic quantities. When all of the dimensions (or units) in an equation balance, the equation is said to be *dimensionally homogeneous.*

Example An alloy has a specific weight of 400 lb$_f$/ft^3. What is the weight of 2 ft^3 of the alloy? Show the numerical and dimensional solutions to the problem.

$$W = V\rho \qquad \text{[Algebraic equation]}$$

or

(Weight[2] of metal) = (volume of metal)(specific weight of metal)

Fundamental Dimensions: $F = (L^3)\left(\frac{F}{L^3}\right)$ [Dimensional equation]

[2]Weight is expressed in the dimensions of force.

Units: $F = (2\ \text{ft}^3)\left(400 \dfrac{\text{lb}_f}{\text{ft}^3}\right) = \mathbf{800\ lb}_f$ [Unit equation]

Check: $\text{lb}_f = \text{lb}_f$

Frequently it will be necessary to change unit systems, that is, feet to inches, hours to seconds, pounds to grams, and so on. This process can be accomplished by the use of unity conversion factors that are multiplied by the expression to be changed. Refer to Appendix IV for a listing of commonly used conversion factors.

Example Change a speed of 3000 miles per hour (miles/hr) to feet per second (ft/sec).

Fundamental dimensions: $\dfrac{L}{T} = \dfrac{L}{T}$

Units: $V = \left(3000 \dfrac{\text{mi}}{\text{hr}}\right)\left(\dfrac{5280\ \text{ft}}{1\ \text{mi}}\right)\left(\dfrac{1\ \text{hr}}{3600\ \text{sec}}\right) = \mathbf{4400 \dfrac{ft}{sec}}$

The two conversion factors, (5280 ft/1 mi) and (1 hr/3600 sec), are each equivalent to unity, since the numerator of each fraction is equal to its denominator (5280 ft = 1 mi, and 1 hr = 3600 sec).

Note that the word *per* means *divided by.* To avoid misunderstandings in computations, the units should be expressed in fractional form.

Example

a. $(X \text{ per } Y) \text{ per } Z = (X \div Y) \div Z = [(X/Y)/Z] = \dfrac{(X/Y)}{Z} = \dfrac{X}{YZ}$

b. Acceleration = 156 ft per sec per min = 156 ft/sec/min

$$= \mathbf{156 \frac{ft}{(sec)(min)}}$$

c. Pressure = 65.4 newtons per square centimetre = $\mathbf{65.4 \dfrac{N}{cm^2}}$

Example Solve for the fundamental dimensions of Q and P in the following dimensionally homogeneous equation if C is a velocity and B is an area.

$$Q = C(B - P)$$

Fundamental Dimensions: $Q = \dfrac{L}{T}(L^2 - P)$

Since the equation is dimensionally homogeneous, P must also be length squared (L^2) in order that the subtraction can be carried out. If this is true, the units of Q are[3]

$$Q = \frac{L}{T}(L^2 - L^2) = \frac{L}{T}(L^2) = \frac{\mathbf{L}^3}{\mathbf{T}}$$

Example Solve for the conversion factor k:

a. $\dfrac{L^2 T^3 \theta}{F^4} = k \left(\dfrac{L^5 TF^2}{Q^2}\right)$

Solving for k: $$k = \frac{T^2 \theta Q^2}{F^6 L^3}$$

and $$\left(\frac{L^2 T^3 \theta}{F^4}\right) = \left(\frac{T^2 \theta Q^2}{F^6 L^3}\right) \left(\frac{L^5 TF^2}{Q^2}\right)$$

Check: $$\frac{L^2 T^3 \theta}{F^4} = \frac{L^2 T^3 \theta}{F^4} = L^2 T^3 \theta F^{-4}$$

b. $\left(\dfrac{F^3 T^2 \theta}{L^2 Q}\right) k = \left(\dfrac{MF^2 Q^3}{T^2 L^3}\right)$

$$k = \frac{\boldsymbol{MQ^4}}{\boldsymbol{FT^4 L\theta}}$$

Check: $$\left[\frac{F^3 T^2 \theta}{L^2 Q}\right]\left[\frac{MQ^4}{FT^4 L\theta}\right] = \frac{MF^2 Q^3}{T^2 L^3} \quad \text{or} \quad MF^2 Q^3 T^{-2} L^{-3}$$

UNITS

The most commonly used fundamental and derived units in engineering calculations are the following:

Units of length The concept of *length* as a measure of space in one direction is easily understood. People in every country use this concept because the position of any point in our universe may be described in relation to any other point by specifying three lengths. The world standard of length is the metre (m), defined now in terms of the wavelength of a particularly uniform monochromatic light. It is quite close to being equal to the distance from the earth's equator to the North Pole divided by ten million, which was its original definition. This unit of length is commonly used by engineers and scientists in most countries for the usual engineering problems as well as in the field of space mechanics.

In the United States, the most common units of length that are used in engineering calculations are the inch (in.), the foot (ft), and the mile (mi).

[3]Remember that the terms L^2 represent a particular length squared in each instance. Thus the remainder (depending on the numerical magnitude of each term) will also be length squared or will be zero for the special case of the original lengths being equal.

Less common are the yard and the nautical mile. They are defined as

$$\begin{aligned} 1 \text{ in.} &= 2.54\,(10)^{-2} \text{ m (exactly by definition)} \\ 1 \text{ ft} &= 12 \text{ in.} \\ 1 \text{ yard} &= 3 \text{ ft} \\ 1 \text{ mile} &= 5280 \text{ ft} \\ 1 \text{ nautical mile} &= 6080.27 \text{ ft approximately} \end{aligned}$$

Often feet and inches or feet and miles are used in the same problem. The foot is sometimes decimalized and sometimes the last fractional foot is given in inches and fractions of an inch. Sometimes the inch is decimalized and sometimes it is fractionalized; sometimes it is both decimalized and fractionalized in the same problem. One of the disadvantages of the English system of measurement is the tendency to use a mixture of methods of showing a measured quantity.

Units of force Force is most commonly thought of as a "push" or a "pull" and represents the action of one body on another. The action may be exerted by direct contact between the bodies or at a distance, as in the case of magnetic and gravitational forces. Weight denotes a quantity of the same nature as a force.

The most common unit of force used by American engineers in the past is the pound (lb_f). This is the force that is required to accelerate a pound mass with the mean acceleration of gravity, or $g = \mathbf{32.174}$ ft/sec^2. Its value is

$$1 \text{ lb}_f = \mathbf{4.48} \text{ newton}$$

The newton (N) is derived from the kilogram by means of Newton's law, $F = Ma$. Thus,

$$1\ N = 1 \text{ kg}_m \times 1 \text{ m/sec}^2$$

and it is the force required to accelerate a 1-kg mass 1 m/sec^2. This unit of force is most frequently used by the American electrical engineers and by other engineers who are engaged in space exploration activities.

Units of time Time cannot be defined in simple terms, but in general it is a measure of the interval separating the occurrence of two events. The mean solar day is the standard unit of time in all systems of units used at present. The hour (hr), the minute (min), and the second (sec) are all derived from the mean solar day, but not decimally. Since all four of these units are used, sometimes even in the same problem, it is easy to see how mistakes can be made and considerable extra work required. This is also the case with the International System of Units (SI), since the unit of time is the same in all

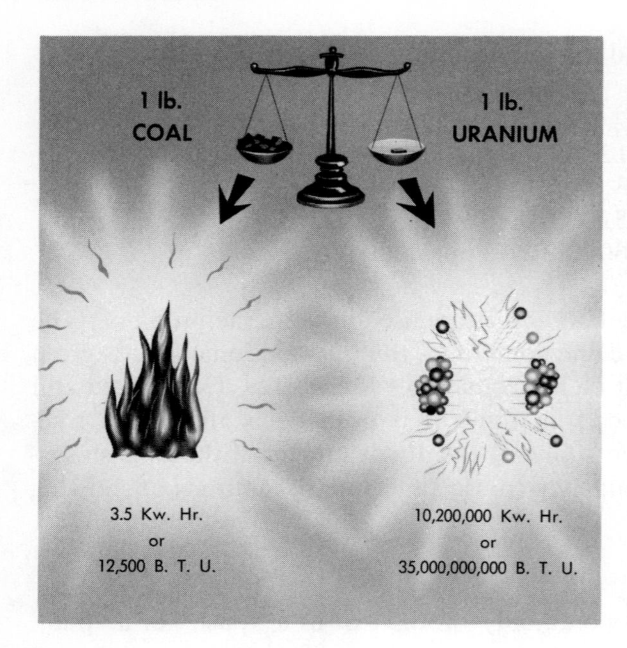

Illustration 4–4. The contrast of energy sources is depicted in this picture, which shows the relative engergy conversion from coal and nuclear fuels. In working with such energy sources, the engineer must be able to convert from one unit system to another. (Courtesy Westinghouse Electric Corporation.)

systems of units. The most common unit used in engineering calculations is the second, defined as

$$1 \text{ sec} = \frac{1 \text{ mean solar day}}{86{,}400}$$

Then, 1 min = 60 sec and 1 hr = 3600 sec.

Units of mass While length, force, and time are readily understood concepts, *mass* is somewhat more difficult to perceive. The universe is filled with matter—the accumulation of electrons, protons, and neutrons. *Mass* is a measure of the quantity of these subatomic particles that a particular object possesses. Although a quantity of matter can change form—for example, as when a block of ice is melted to water and then vaporized to steam—its "quantity" does not change.

In contrast to length, force, and time, there is no direct measure for mass. Its quantity may be measured only through an examination of its properties, such as the amount of force that must be provided to give it a certain acceleration. The world standard of mass is the kilogram (kg), defined originally as being one thousandth of 1 m^3 of water at a temperature of 4° Celsius and standard atmospheric pressure, but now defined as the mass of a block of platinum kept at the French Bureau of Standards. This unit is used by American electrical and space engineers.

The pound mass (lb_m) is the unit that the average American engineer thinks he is using most of the time. In most instances this is incorrect.

Generally the pound that he uses is the pound force (lb_f), from which he derives units of mass by means of Newton's law, $F = Ma$. Thus,

$$1 \text{ lb}_m = 0.4535924277 \text{ kg (by definition)}$$

$$1 \frac{\text{lb}_f \text{ sec}^2}{\text{ft}} \text{ (also called a slug)} = 32.174 \text{ lb}_m$$

$$1 \frac{\text{lb}_f \text{sec}^2}{\text{in.}} = 386.088 \text{ lb}_m$$

Units of temperature Temperature is an arbitrary measure which is proportional to the average kinetic energy of the molecules of an ideal gas. Four temperature scales are used by American engineers. The degree Celsius (°C), formerly called centigrade, reads zero at the freezing point of water and 100°C at the boiling point of water under standard conditions of pressure. It is the world standard of temperature. The temperature in degrees Kelvin (°K) is derived from the Celsius scale by the following equation:

$$\text{Temperature in °K} = \text{Temperature in °C} + 273.16$$

The degree Fahrenheit (°F) reads 32°F at the freezing point of water and 212°F at the boiling point of water under standard conditions of pressure.[4] The degree Rankine (°R) is derived from the Fahrenheit scale by the equation

$$\text{Temperature in °R} = \text{Temperature in °F} + 459.69$$

When a temperature is measured in either °K or in °R, it is said to be the *absolute* temperature, because these scales read zero for the condition where the kinetic energy of the molecules of an ideal gas is presumed to be zero.

Units of area and volume Units of area and volume are derived from the units of length for the most part. However, the gallon is a commonly used measure of volume that is in no way related to units of length.

Units of velocity and acceleration These units are all derived from the fundamental units of length and time.

Units of work and energy Work is the product of a force and a distance through which that force acts. Energy is the ability or capacity for doing

[4]G. D. Fahrenheit thought that 0°F was the lowest possible temperature that could be obtained and that 100°F was the uniformly standard temperature of human blood.

work. Although the two quantities are conceptually different, they are measured by the same units. Several different units are used. For example, foot-pounds, inch-pounds, and horsepower-hours are all used for work and mechanical energy; both the joule and the kilowatt-hour are used for electrical energy; and both the calorie (two types) and the British thermal unit (Btu) are used for heat energy. In some problems all of these units occur, which frequently makes the task of unit conversion the most formidable part of the solution.

Units of power Power is the time rate of accomplishing work. The average power is the work performed divided by the time required for the performance. Since power units are derived units, they also involve the various work and energy units described above. In addition, the ton of refrigeration (3517 watts) is sometimes used in air conditioning design calculations.

Units of pressure Pressure is the result of a force distributed over an area. In general, the units of pressure have been derived from conventional units of force and area. However, other measures are also used. For example, the standard atmospheric pressure is commonly used as a unit, and for fractional atmospheres the millimeter of mercury, the inch of mercury, and the inch of water are used.

Units of density Density is a measure of the mass that a body of uniform substance possesses per unit volume. The units of density are derived units that are made up by dividing the chosen unit of *mass* by the unit of volume, for example, grams per cubic centimeter.

Units of specific weight Specific weight is a measure of the *weight* of a substance per unit volume. Many people confuse density with specific weight. Remember that they are not the same, but rather that they are related to each other by the relationship.

$$(\text{Specific Weight}) = (\text{Density})(\text{Acceleration of Gravity})$$

The units most commonly used for specific weight are lb/ft^3 and lb/in.^3, where the lb is a unit of force, lb_f. They represent the attraction, in lb_f, of the earth on either 1 ft^3 or 1 in.^3 of the material. To convert specific weights to density for use in a gravitational system, one must divide these quantities by the acceleration of gravity, g. If one divides the first unit by $g = 32.174$ ft/sec², one obtains the unit of density $lb_f sec^2/ft^4$, or slugs/ft³, and, if one divides the second unit by $g = 386.088$ in./sec², one obtains the unity of density $lb_f sec^2/in^4$.

Problems

Solve for the conversion factor k.

4-1. $k = \left(\frac{L^3\theta TQ^5}{FM}\right) = \frac{M^3\theta Q}{L^2}$

4-2. $\frac{FTL^2}{\theta M^3} = k\frac{\theta^5 M}{T^2}$

4-3. $k\left(\frac{QM}{TF^2}\right) = \sqrt{L^4 I\theta Q^8}$

4-4. $\theta^2\sqrt{LM^5} = k\left(\frac{FT^2}{M^3}\right)$

4-5. $k(F\theta^2 TL^{-2}M^{-3}) = M^5 L\theta F^{-3}$

4-6. $M^2FT^{-5}L^{-2} = k\sqrt{MT\theta}$

4-7. $\sqrt{LT^3F^{-2}M} = k\sqrt{TF^3M^6}$

4-8. $k\frac{\sqrt{T^3Q}}{L^2F^{-2}} = MTLF$

4-9. $k(F^2T\sqrt{L\theta^{-2}}) = \theta^{-3}T^{-2}$

4-10. $FL^3Q^{-1}M^{-3} = k\sqrt{L^2Q^{-1}}$

4-11. Convert 76 newtons to dynes and lb_f.

4-12. Convert 2.67 in. to angstroms and miles.

4-13. Convert 26 knots to feet per second and meters per hour.

4-14. Convert 8.07(10)³ tons to grams and ounces.

4-15. Convert 1.075 atmospheres to dynes per cm² and inches of mercury.

4-16. Convert 596 Btu to foot-pounds and Joules.

4-17. Convert 26,059 watts to horsepower and ergs per second.

4-18. Convert 92.7 coulombs to faradays.

4-19. Convert 75 angstroms to feet.

4-20. Convert 0.344 henries to abhenries.

4-21. Express 2903 ft³ of sulphuric acid in gallons and cubic meters.

4-22. Change a Btu to horsepower-seconds.

4-23. A car is traveling 49 mi/hr. What is the speed in feet per second and meters per second?

4-24. A river has a flow of 3(10)⁶ gallons per 24-hour day. Compute the flow in cubic feet per minute.

4-25. Convert 579 qt/sec to cubic feet per hour and cubic meters per second.

4-26. A copper wire is 0.0809 cm in diameter. What is the weight of 1000 m of the wire?

4-27. A cylindrical tank 2.96 ft high has a volume of 136 gallons. What is its diameter?

4-28. A round iron rod is 0.125 in. in diameter. How long will a piece have to be to weigh 1 lb?

4-29. Find the weight of a common brick that is 2.6 in. by 4 in. by 8.75 in.

4-30. Convert 1 yd² to acres and square meters.

4-31. A white pine board is 14 ft long and 2 in. by 8 in. in cross section. How much will the board weigh? At $200.00 per 1000 f.b.m., what is its value?

4-32. A container is 12 in. high, 10 in. in diameter at the top, and 6 in. in diameter at the bottom. What is the volume of this container in cubic inches? What is the weight of mercury that would fill this container?

4-33. How many gallons of water will be contained in a horizontal pipe 10 in. in internal diameter and 15 ft long, if the water is 6 in. deep in the pipe?

4-34. A hemispherical container 3 ft in diameter has half of its volume filled with lubricating oil. Neglecting the weight of the container, how much would the contents weigh if kerosene were added to fill the container to the brim?

4-35. What is the cross-sectional area of a railroad rail 33 m long that weighs 94 lb/yd?

4-36. A piece of cast iron has a very irregular shape and its volume is to be determined. It is submerged in water in a cylindrical tank having a diameter of 16 in. The water level is raised 3.4 in. above its original level. How many cubic feet are in the piece of cast iron? How much does it weigh?

4-37. A cylindrical tank is 22 ft in diameter and 8 ft high. How long will it take to fill the tank with water from a pipe which is flowing at 33.3 gallons/min?

4-38. Two objects are made of the same material and have the same weights and diameters. One of the objects is a sphere 2 m in diameter. If the other object is a right cylinder, what is its length?

4-39. A hemisphere and cone are carved out of the same material and their weights are equal. The height of the cone is 3 ft, $10\frac{1}{2}$ inches while the radius of the hemisphere is 13 in. If a flat circular cover were to be made for the cone base, what would be its area in square inches?

4-40. An eight-sided wrought iron bar weighs 3.83 lb per linear foot. What will be its dimension across diagonally opposite corners?

4-41. Is the equation $a = (2S/t^2) - (2V_1/t)$ dimensionally homogeneous if a is an acceleration, V_1 is a velocity, t is a time, and S is a distance? Prove your answer by writing the equation with fundamental dimensions.

4-42. Is the equation $V_2^2 = V_1^2 + 2as$ dimensionally correct if V_1 and V_2 are velocities, a is an acceleration, and s is a distance? Prove your answer by rewriting the equation in fundamental dimensions.

4-43. In the homogeneous equation $R = B + \frac{1}{2}CX$, what are the fundamental dimensions of R and B if C is an acceleration and X is a time?

4-44. Determine the fundamental dimensions of the expression $B/g\sqrt{D - m^2}$, where B is a force, m is a length, D is an area, and g is the acceleration of gravity at a particular location.

4-45. The relationship $M = \sigma I/c$ pertains to the bending moment for a beam under compressive stress. σ is a stress in F/L^2, C is a length L, and I is a moment of inertia L^4. What are the fundamental dimensions of M?

4-46. The expression $V/K = (B - \frac{7}{3}A)A^{5/3}$ is dimensionally homogeneous. A is a length and V is a volume of flow per unit of time. Solve for the fundamental dimensions of K and B.

4-47. Is the expression $S = 0.031V^2/fB$ dimensionally homogeneous if S is a distance, V is a velocity, f is the coefficient of friction, and B is a ratio of two weights? Is it possible that the numerical value 0.031 has fundamental dimensions? Prove your solution.

4-48. If the following heat transfer equation is dimensionally homogeneous, what are the units of k?

$$Q = \frac{-kA(T_1 - T_2)}{L}$$

A is a cross-sectional area in square feet, L is a length in feet, T_1 and T_2 are temperatures (°F), and Q is the amount of heat (energy) conducted in Btu per unit of time.

4-49. In the dimensionally homogeneous equation

$$F = \frac{4Ey}{(1-\mu^2)(Md^2)}\left[(h-y)\left(h-\frac{y}{2}\right)t - t^3\right]$$

F is a force, E is a force per (length)^2, y, d, and h are lengths, μ is Poisson's ratio, and M is a ratio of diameters. What are the fundamental dimensions of t?

UNIT SYSTEMS

Unit systems are of two general types—absolute and gravitational. The absolute systems are independent of gravitational effects on the earth or other planets and are generally used for scientific calculations. In absolute systems the dimensions of force are derived in terms of the fundamental units of time, length, and mass. There are three absolute systems. Two of these are used extensively in scientific work today. These are the SI (metre, kilogram, second) absolute system, and the CGS (centimeter, gram, second) absolute system. The other absolute system, the FPS absolute system, has been used primarily in engineering computations. The American Engineering is the more commonly used of the gravitational systems.

Table 4–1 Unit Systems

	Absolute			*Gravitational*	
	(1) SI (Modified MKS)	*(2) CGS*	*(3) FPS*	*(4) FPS*	*(5) American Engineering*
Fundamental dimensions					
Force (F)	—	—	—	lb_f	lb_f
Length (L)	m	cm	ft	ft	ft or in.
Time (T)	sec	sec	sec	sec	sec
Mass (M)	kg_m	g	lb_m	—	lb_m
Derived dimensions					
Force (F)	$\frac{kg_m\text{-m}}{sec^2}$ (called a *newton**)	$\frac{\text{g-cm}}{sec^2}$ (called a *dyne*)	$\frac{lb_m\text{-ft}}{sec^2}$ (called a *poundal*)	—	—
Mass (M)	—	—	—	$\frac{lb_f\text{-}sec^2}{ft}$ (called a *slug*)	—

Table 4–1 (cont)

	Absolute			*Gravitational*	
	(1) SI (Modified MKS)	*(2) CGS*	*(3) FPS*	*(4) FPS*	*(5) American Engineering*
Energy (LF)	N-m	cm-dyne (called an *erg*)	ft-poundal	ft-lb_f	ft-lb_f
Power $\left(\frac{LF}{T}\right)$	$\frac{\text{N-m}}{\text{sec}}$	$\frac{\text{erg}}{\text{sec}}$	$\frac{\text{ft-poundal}}{\text{sec}}$	$\frac{\text{ft-lb}_f}{\text{sec}}$	$\frac{\text{ft-lb}_f}{\text{sec}}$
Velocity $\left(\frac{L}{T}\right)$	$\frac{\text{m}}{\text{sec}}$	$\frac{\text{cm}}{\text{sec}}$	$\frac{\text{ft}}{\text{sec}}$	$\frac{\text{ft}}{\text{sec}}$	$\frac{\text{ft}}{\text{sec}}$
Acceleration $\left(\frac{L}{T^2}\right)$	$\frac{\text{m}}{\text{sec}^2}$	$\frac{\text{cm}}{\text{sec}^2}$	$\frac{\text{ft}}{\text{sec}^2}$	$\frac{\text{ft}}{\text{sec}^2}$	$\frac{\text{ft}}{\text{sec}^2}$
Area (L^2)	m^2	cm^2	ft^2	ft^2	ft^2
Volume (L^3)	m^3	cm^3	ft^3	ft^3	ft^3
Density $\left(\frac{M}{L^3}\right)$	$\frac{\text{kg}_m}{\text{m}^3}$	$\frac{\text{g}}{\text{cm}^3}$	$\frac{\text{lb}_m}{\text{ft}^3}$	$\frac{\text{lb}_f\text{sec}^2}{\text{ft}^4}$	$\frac{\text{lb}_m}{\text{ft}^3}$
Pressure $\left(\frac{F}{L^2}\right)$	$\frac{\text{N}}{\text{m}^2}$	$\frac{\text{dyne}}{\text{cm}^2}$	$\frac{\text{poundal}}{\text{cm}^2}$	$\frac{\text{lb}_f}{\text{ft}^2}$	$\frac{\text{lb}_f}{\text{ft}^2}$

**A newton (N) is the force required to accelerate a 1-kg mass at 1 m/sec².* The acceleration of gravity at sea level and 45° latitude has the measured value of 9.807 m/sec². A force of 1 kg equals 9.807 N of force.

THE METRIC (SI) SYSTEM

The majority of the countries of the world use the metric system (also called the MKS–metre, kilogram, second–system). This system was initiated in France in 1790, during the French Revolution, when the National Assembly of France requested the French Academy of Sciences to "deduce an invariable standard for all the measures and all the weights." The unit of length was to be a fraction of the earth's circumference, with other measures (such as volume and mass) to be derived from it. Also, the various sized parts of each unit were to be a multiple of ten of that unit, making it a decimal system.

France made the use of the metric system compulsory in 1840, and its use by other countries has increased steadily since that time. In 1866 the United States Congress passed a statute making it "lawful throughout the United States of America to employ the weights and measures of the metric system in all contracts, dealings or court proceedings." However, it was not until the recent era of space travel that its use became widely used in scientific work. Most of the transactions of commerce still use the English FPS (foot, pound, second) gravitational system in spite of its inherent conversion complexities. However, industry is converting rapidly to the metric system.

In 1960, the General Conference of Weights and Measures–formed of the international adherents to the metric convention–adopted an extensive revision and simplification of the metric system. The name *Le Système International d'Unités* (International System of Units), with the abbreviation SI, was adopted. Further improvements were made in 1964, 1968, and 1971. It is anticipated that within a few years all transactions in this country will use the SI system of units.

The SI (metric) system is particularly convenient in calculations involving energy, since only one unit is used for all types of energy, whether atomic, electric, chemical, heat or mechanical. This unit of energy is the joule, which previously was used only by electrical engineers, and the corresponding unit of power is the watt, which is 1 joule/sec.

The following is a complete list of SI units and their dimensions:

SI Fundamental units (also called base units)

Length: 1 m (metre)
Mass: 1 kg_m (kilogram)
Time: 1 s (second)
Electric Current = 1 ampere (amp) = 1 coulomb/sec
Thermodynamic Temperature: 1°K (kelvin)
Luminous Intensity: 1 candela (International candle)
Amount of Substance: 1 mole

Examples of Derived units

area = 1 m^2
volume = 1m^3
velocity = 1 m/sec
acceleration = 1 m/sec^2
force = 1 newton (N) = 1 kg_m-m/sec^2
work and energy: 1 joule (j) = 1 kg_m-m^2/sec^2
moment and torque = 1 N-m
power, 1 watt (w) = 1 kg_m-m^2/sec^3
pressure, 1 N/m^2 = 1 kg_m/sec^2-m (1 bar = 10^5 N/m)
thermal conductivity = 1 w/m-°C = 1 kg_m-m/sec^3-°C
heat transfer coefficient = 1 w/m^2-°C = 1 kg_m/sec^3-°C
dynamic viscosity = 1 N-sec/m^2 = 1 kg_m/m-sec = 1 decapoise
kinematic viscosity = 1 m^2/sec = 1 myriastoke
density = 1 kg_m/m^3
heat coefficient = 1 j/kg_m-°C = 1 m^2/sec^2-°C
enthalpy, heat content, and internal energy = 1 j/kg_m =
1 m^2/sec^2

electrical charge; 1 coulomb = 1 amp-sec
potential = 1 volt (v) = 1 w/amp = 1 kg_m-m^2/sec^3-amp
resistance = 1 ohm (Ω) = 1 w/amp^2 = 1 kg_m-m^2/sec^3-amp^2
capacitance = 1 farad (f) = 1 coulomb/v = 1 amp^2-sec/w =
1 amp^2-sec^4/kg_m m^2
inductance = 1 henry (h) = 1 v-sec/amp = 1 j/amp^2 =
1 kg_m-m^2/sec^2-amp^2
capacity or permittivity $\epsilon_o = \frac{10^7}{4\pi c^2} = 8.854 \times 10^{-12}$ f/m
magnetic permeability $\mu_0 = 4\pi(10)^{-7} = 1.2566(10)^{-6}$ h/m

The American engineering system of units

Early in the development of engineering analysis a system of units was developed that defined both the units of mass and the units of force. It is perhaps unfortunate that the same word, pounds, was chosen to represent both quantities, since they are physically different. In order to help differentiate the quantities, the pound-mass may be designated as lb_{mass} (or lb_m) and the pound-force as lb_{force} (or lb_f).

For many engineering applications the numerical values of lb_m and lb_f are very nearly the same. However, in expressions such as $F = Ma$, it is necessary that the difference between lb_m and lb_f be maintained. By definition, a mass of 1 lb_m will be attracted to the earth by a force of 1 lb_f at a place where the acceleration of gravity is 32.2 ft/sec^2. If the acceleration of gravity changes to some other value, the force must change in proportion, since mass is invariant.

Although the pound subscripts, *force* and *mass,* are frequently omitted in engineering and scientific literature, it is nevertheless true that lb_f is not the same as lb_m. Their numerical values are equal, however, in the case of sea level, 45°-latitude calculations. However, their values may be widely different, as would be the case in an analysis involving satellite design and space travel.

In Newton's equation, $F = Ma$, dimensional homogeneity must be maintained. If length, force, and time are taken as fundamental dimensions, the dimensions of mass must be derived. This can be accomplished as follows:

$$F = Ma$$

Then

$$M = \frac{F}{a}$$

$$= \frac{(F)}{(L/T^2)} = \frac{FT^2}{L} = FL^{-1}T^2$$

and

$$= \mathbf{lb_f\text{-}ft^{-1}\text{-}sec^2}$$

For convenience, this derived unit of mass (1 lb-sec^2)/ft is called a *slug.* Thus, a force of 1 lb_f will cause a mass of 1 slug to have an acceleration of 1 ft/sec^2.

The relationship between 1 lb_m and 1 slug is given by considering that whereas 1 lb_f will accelerate 1 lb_m with an acceleration of $g = 32.2$ ft/sec^2, it will accelerate 1 slug with an acceleration of only 1 ft/sec^2. Thus:

$$1\ lb_f = (1\ lb_m)(32.2\ ft/sec^2) = (1\ slug)(1\ ft/sec^2)$$

or

$$1\ slug = 3.22\ lb_m$$

It should be noted that with the FPS system a unity conversion factor must be used if a mass unit other than the slug is used. Since the acceleration of gravity varies with both latitude and altitude, the use of a gravitational system is sometimes inconvenient. A 100,000-lb rocket on the earth, for example, would not weigh 100,000 lb_f on the moon, where gravitational forces are smaller. The mass of the rocket, on the other hand, is a fixed quantity and will be a constant amount, regardless of its location in space.

For a freely falling body at sea level and 45° latitude, the acceleration[5] g of the body is 32.174 (approximately 32.2) ft/sec^2. As the mass is attracted to the earth, the only force then acting on it is its own weight.

then $F = Ma$

If $W = Mg$

and $M = \dfrac{W}{g}$

where $a = g$ and $F = W$

In this particular system of units, then, the mass of a body in slugs may be calculated by dividing the weight of the body in pounds by the local acceleration of gravity in feet per second squared.

The engineer frequently works in several systems of units in the same calculation. In this case it is only necessary that the force, mass, and acceleration dimensions all be expressed in any valid set of units from any unit system. Numerical equality and unit homogeneity may be determined in any case by applying unity conversion factors to the individual terms of the expression.

Example Solve for the lb_m which is being accelerated at 3.07 ft/sec^2 by a force of 392 lb_f.

[5] The value of the acceleration of gravity, g, at any latitude θ on the earth may be approximated from the following relationship: $g = 32.09(1 + 0.0053 \sin^2 \theta)$ ft/sec^2.

Solution $F = Ma$ or $M = \dfrac{F}{a}$

$$M = \frac{392\ \text{lb}_f}{3.07\ \text{ft/sec}^2} = \mathbf{127.8}\,\frac{\mathbf{lb}_f\textbf{-sec}^2}{\mathbf{ft}}$$

The direct substitution has given mass in the units of slugs instead of lb_m units. This is a perfectly proper set of units for mass, although not in lb_m units as desired. Consequently the final equation must be altered by applying the unity conversion factor

$$\left(\frac{32.2\ \text{lb}_m}{1\ \text{lb}_f\text{-sec}^2\text{-ft}^{-1}}\right).$$

The object, of course, is to cancel units until the desired units appear in the answer. Thus

$$M = \left(\frac{127.8\ \text{lb}_f\text{-sec}^2}{\text{ft}}\right)\left(\frac{32.2\ \text{lb}_m\text{-ft}}{1\ \text{lb}_f\text{-sec}^2}\right) = \mathbf{(4.11)(10)^3\ lb}_m$$

Example Solve for the mass in slugs being accelerated at 13.6 m/sec by a force of 1782 lb_f.

Solution $F = Ma$

$$M = \frac{F}{a} = \frac{(1782\ \text{lb}_f)}{(13.6\ \text{m/sec}^2)} = \left(\frac{1782\ \text{lb}_f\text{-sec}^2}{13.6\ \text{m}}\right)\left(\frac{1\ \ \text{m}}{3.28\ \text{ft}}\right)$$

$$= 40\,\frac{\text{lb}_f\text{-sec}^2}{\text{ft}} = \mathbf{40\ slugs}$$

It is recommended that in writing a mathematical expression to represent some physical phenomena, the engineer should avoid using stereotyped conversion symbols such as g, g_c, k, or J in the equation. If one of these, or any other conversion factor, is needed in an equation to achieve unit balance, it can *then* be added. Since many different unit systems may be used from time to time, it is best to add unity conversion factors *only* as they are needed. Unfortunately, in much engineering literature, the equations used in a particular instance have been written to include one or more unity conversion factors. Considerable care must be exercised, therefore, in using these expressions since they represent a "special case" rather than a "general condition." The engineer should form a habit of always checking the unit balance of all equations.

Remember that

$$1\ \text{slug} = 1\,\frac{\text{lb}_f\text{-sec}^2}{\text{ft}} = 32.2\ \text{lb}_m$$

The foregoing discussion has shown that

1. If mass units in slugs are used in the expression $F = Ma$, the force units will come out in the usual units of pounds (lb_f).
2. If mass units in pounds (lb_m) are used in the expression $F = Ma$, force units will come out in an absolute unit called the *poundal* (see Table 4–1).

In engineering calculations the inch is used just as often as the foot to represent the unit of length, and this necessitates the introduction of an additional unit of mass. Consider Newton's law, $F = Ma$, where $F = 1\ lb_f$ and $a = 1$ in./sec². Then

$$M = 1\ \text{lb}_f\text{-sec}^2/\text{in.}$$

where lb_f now is lb_{force}.

Example A body weighs $W\ lb_f$ at a place where $g = 386$ in./sec². Find the mass of the body in units of lb_f-sec²/in.

Solution The relationship between weight and mass is given by

$$W = Mg$$

and if W is given in lb_f and g in in./sec², this gives

$$M = \frac{W}{g} = \frac{W}{386}\ \text{lb}_f\text{-sec}^2/\text{in.}$$

Problems on unit systems

4–50. The kinetic energy of a moving body in space can be expressed as follows:

$$KE = \frac{MV^2}{2}$$

where KE = kinetic energy of the moving body
M = mass of the moving body
V = velocity of the moving body

a. Given: $M = 539 \dfrac{\text{lb}_f\text{-sec}^2}{\text{ft}}$; $V = 2900 \dfrac{\text{ft}}{\text{sec}}$

Find: KE in ft-lb_f

b. Given: $M = 42.6 \dfrac{\text{lb}_f\text{-sec}^2}{\text{ft}}$; $KE = 1.20(10)^{11}$ ft-lb_f

Find: V in $\dfrac{\text{ft}}{\text{sec}}$

c. Given: $KE = 16{,}900$ in.-lb_f; $V = 3960 \frac{\text{in.}}{\text{min}}$

Find: M in slugs

d. Given: $M = 143$ g; $KE = 2690$ in.-lb_f

Find: V in $\frac{\text{mi}}{\text{hr}}$

4-51. The inertia force due to the acceleration of a rocket can be expressed as follows:

$$F = Ma$$

where F = unbalanced force
a = acceleration of the body
M = mass of the body

a. Given: $a = 439 \frac{\text{ft}}{\text{sec}^2}$; $M = 89.6 \frac{lb_f\text{-sec}^2}{\text{ft}}$

Find: F in lb_f

b. Given: $F = 1500$ lb_f; $M = 26.4 \frac{lb_f\text{-sec}^2}{\text{ft}}$

Find: a in $\frac{\text{ft}}{\text{sec}^2}$

c. Given: $F = (49.3)(10)^5$ lb_f; $a = 32.2 \frac{\text{ft}}{\text{sec}^2}$

Find: M in $\frac{lb_f\text{-sec}^2}{\text{ft}}$

d. Given: $M = 9650 \frac{lb_f\text{-sec}^2}{\text{ft}}$; $a = 980 \frac{\text{cm}}{\text{sec}^2}$

Find: F in lb_f

4-52. The force required to assemble a force-fit joint on a particular piece of machinery may be expressed by the following equation:

$$F = \frac{\pi dlfP}{2000}$$

where d = shaft diameter, in.
l = hub length, in.
f = coefficient of friction
P = radial pressure, psi
F = force of press required, tons

a. Given: $d = 9.05$ in.; $l = 15.1$ in.; $f = 0.10$; $P = 10{,}250$ psi
Find: F in lb_f

b. Given: $F = 4.21 \times 10^5$ lb_f; $f = 0.162$; $P = 8.32(10^8)$ psf; $l = 1.62$ ft
Find: d in ft

c. Given: $d = 25$ cm; $l = 30.2$ cm; $f = 0.08$; $P = 9260$ psi
Find: F in tons

d. Given: $F = 206$ tons; $d = 6.23$ in.; $l = 20.4$ in.; $f = 0.153$
Find: P in lb_f/ft^2

4-53. The dynamic stress in the rim of a certain flywheel has been expressed by the following equation:

$$\sigma = 0.0000284\rho r^2 n^2$$

where σ = tensile stress, $\frac{\text{lb}_f}{\text{in.}^2}$

ρ = specific weight of material, $\frac{\text{lb}_f}{\text{in.}^3}$

r = radius of curvature, in.
n = number of rpm

a. Given: $\sigma = 200$ psi; $\rho = 0.282 \frac{\text{lb}_f}{\text{in.}^3}$; $r = 9$ in.

Find: n in rpm

b. Given: $\rho = 0.332 \frac{\text{lb}_f}{\text{in.}^3}$; $r = 23.1$ cm; $m = 200$ rpm

Find: σ in psi

4-54. Assuming that the acceleration due to gravitation is 5.31 fps^2 on the moon, what is the mass in slugs of 100 lb_m located on the moon?

4-55. A silver bar weighs 382 lb_f at a point on the earth where the acceleration of gravity is measured to be 32.1 fps^2. Calculate the mass of the bar in lb_m and slug units.

4-56. The acceleration of gravity can be approximated by the following relationship:

$$g = 980.6 - (3.086)(10)^{-6}A$$

where g is expressed in cm/sec^2, and A is an altitude in cm. If a rocket weighs 10,370 lb_f at sea level and standard conditions, what will be its weight in dynes at 50,000-ft elevation?

4-57. At a certain point on the moon the acceleration due to gravitation is 5.35 fps^2. A rocket resting on the moon's surface at this point weighs 23,500 lb_f. What is its mass in slugs? In lb_m?

4-58. If a 10-lb weight on the moon (where $g = 5.33$ fps^2) is returned to the earth and deposited at a latitude of 90° (see page 244), how much would it weigh in the new location?

4-59. A 4.37-slug mass is taken from the earth to the moon and located at a point where $g = 5.33$ fps^2. What is the magnitude of its mass in the new location?
Is the equation $F = WV^2/2g$ a homogenous expression if W is a weight, V is a velocity, F is a force, and g is the linear acceleration of gravity? Prove your answer, using the FPS absolute system of units.

4-60. Sir Isaac Newton expressed the belief that all particles in space, regardless of their mass, are each attracted to every other particle in space by a specific force of attraction. For spherical bodies, whose separation is very large compared with the physical dimensions of either particle, the force of attraction may be calculated from the relationship $F = Gm_1m_2/d^2$, where F is the existing gravitational force, d is the distance separating the two masses m_1 and m_2, and G is a gravitational constant, whose magnitude depends upon the unit system being used. Using the CGS absolute system of units [$G = 6.67 \times 10^{-8}$ (cm^3/gm-sec^2)], calculate the mass of the earth if it attracts a mass of 1 g with a force of 980 dynes. Assume that the distance from the center of the earth to the gram mass is 6370 km.

4-61. Referring to Problem 4-60, calculate the mass of the sun if the earth (6×10^{24} kg mass) has an orbital diameter of 1.49×10^7 km and the force of attraction between the two celestial bodies is $(1.44)(10)^{25}$ N.

4-62. From Problem 4-60, calculate the acceleration of gravity on the earth in CGS absolute units.

4-63. An interstellar explorer is accelerating uniformly at 58.6 fps^2 in a spherical space ship which has a total mass of 100,000 slugs. What is the force acting on the ship?

•**4-64.** At a certain instant in time a space vehicle is being acted on by a vertically upward thrust of 497,000 lb_f. The mass of the space vehicle is 400,000 lb_m, and the acceleration of gravity is 32.1 fps^2. Is the vehicle rising or descending? What is its acceleration? (Assume "up" means radially outward from the center of the earth.)

4-65. Some interstellar adventurers land their spacecraft on a certain celestial body. Explain how they could calculate the acceleration of gravity at the point where they landed.

4-66. In a swimming pool manufacturer's design handbook, for a pool whose surface area is triangular, you find the following formula: $V = 3.74Rt\theta$, where V = volume of pool in gallons, R = length of base of triangular shaped pool in feet, t = altitude of triangular shaped pool in feet if t is measured perpendicular to R, and θ = average depth of pool in feet. Prove that the equation is valid or invalid.

4-67. You are asked to check the engineering design calculations for a sphere-shaped satellite. At one place in the engineer's calculations you find the expression $A = 0.0872\Delta^2$, where A is the surface area of the satellite measured in square feet, and Δ is the diameter of the satellite measured in inches. Prove that the equation is valid or invalid.

4-68. The U.S. Navy is interested in your torus-shaped lifebelt design and you have been asked to supply some additional calculations. Among these is the request to supply the formula for the volume of the belt in cubic feet if the average diameter of the belt is measured in feet and the diameter of a typical cross-sectional area of the belt is measured in inches. Develop the formula.

4-69. From the window of their spacecraft two astronauts see a satellite with foreign insignia markings. They maneuver for a closer examination. Apparently the satellite has been designed in the shape of an ellipsoid. One of the astronauts quickly estimates its volume in gallons from the relationship $V = 33.8ACE$, where the major radius (A) is measured in meters, the minor radius (E) is measured in feet, and the endview depth to the center of the ellipsoid (C) is measured in centimeters. Verify the correctness of the mathematical relationship used for the calculations.

4-70. An engineer and his family are visiting in Egypt. The tour guide describes in great detail the preciseness of the mathematical relationships used by the early Egyptians in their construction projects. As an example he points out some peculiar indentations in a large stone block. He explains that these particular markings are the resultant calculations of "early day" Egyptians pertaining to the volume of the pyramids. He says that the mathematical relationship used by these engineers was $\odot = \square \uparrow$, where $\odot$ was the volume of pyramid in cubic furlongs, $\square$ was the area of the pyramid base in square leagues, and $\uparrow$ was the height of the pyramid in hectometers. The product of the area and the height equals the volume. The engineer argued that the guide was incorrect in his interpretation. Prove which was correct.

4-71. Develop the mathematical relationship for finding the weight in drams of a truncated cylinder of gold if the diameter of the circular base is measured in centimeters and the height of the piece of precious metal is measured in decimeters.

4-72. If a silver communications satellite has a mass of 126.3 lb_m at Houston, Texas, what would be its weight in newtons on the moon, where the acceleration of gravity is measured to be 162 cm/sec^2?

4-73. A volt is defined as the electric potential existing between two points when 1 joule of work is required to carry 1 coulomb of charge from one point to the other. An ampere is defined as a flow of 1 coulomb of charge per second in a conducting medium. From these definitions, derive an expression for power in watts in an electrical circuit.

4-74. An electric light bulb requires 100 w of power while burning. At what rate is heat being produced? What will be the horsepower corresponding to 100 w?

4-75. A 440-v electric motor which is 83 per cent efficient is delivering 4.20 hp to a hoist which is 76 per cent efficient. At what rate can a mass of 1155 kg be lifted?

4–76. How many kilograms of silver will be transferred in an electroplating tank by a passage of 560 amp for 1 hr? (Hint: 96,500 coulombs will deposit a gram-equivalent of an element in a plating solution.)

4–77. A window-mount type of air conditioning unit is rated at $\frac{3}{4}$ ton capacity for cooling. If the overall efficiency of the motor and compressor unit is 26 per cent, what electric current will be necessary to operate the unit continuously when connected to a 120-v alternating current power line?

4–78. A large capacitor is rated at 10,000 microfarads. If it is connected to a 6.3-v battery, how many coulombs will be required to charge it?

4–79. A capacitor used in transistor circuits is rated at 5 picofarads. How many coulombs will be required to charge it if it is connected to a 9-v battery?

4–80. The reactance in ohms of a coil of wire is given as $X_L = 2fL$, where f is the frequency of an electric current in cycles per second and L is the coil inductance in henries. Compute the reactance of a small solenoid coil whose inductance is 2.75 millihenries if the coil is connected to a 109-v line whose frequency is 412 hertz.

5

the problem solving process

Problem solving may be considered in some degree to be both *art* and *science.* The *art* of problem solving is developed over a period of continuous practice, whereas the *science* of problem solving comes about through a study of the engineering method of problem solving. Both engineers and scientists must be "problem solvers." However, in many instances the end product of the engineer's design, which is a working system economically devised, is considerably different from that of the scientist's, which may be a solution without regard to economics or usefulness.

To many people engineering design means the making of engineering drawings, putting on paper ideas that have been developed by others, and perhaps supervising the construction of a working model. While engineers should possess the capability to do these things, the process of engineering design includes much more: the *formulation* of problems, the *development* of ideas, their *evaluation* through the use of models and analysis, the *testing* of the models, and the *description* of the design and its function in proposals and reports.

An engineering problem may appear in any size or complexity. It may be so small that an engineer can complete it in one day or so large that it will take a team of engineers many years to complete. It may call for the design of a tiny gear in a big machine, perhaps the whole machine, or an entire plant or process which would include the machine as one of its components. When the design project gets so big that its individual components can no longer be stored in one man's head, then special techniques are required to catalog all of the details and to ensure that the components of the system work harmoniously as a coherent unit. The techniques which have been developed to ensure such coordination are called *systems design.*

Regardless of the complexity of a problem that might arise, the *method* for solving it follows a pattern similar to that represented in Figure 5-1.

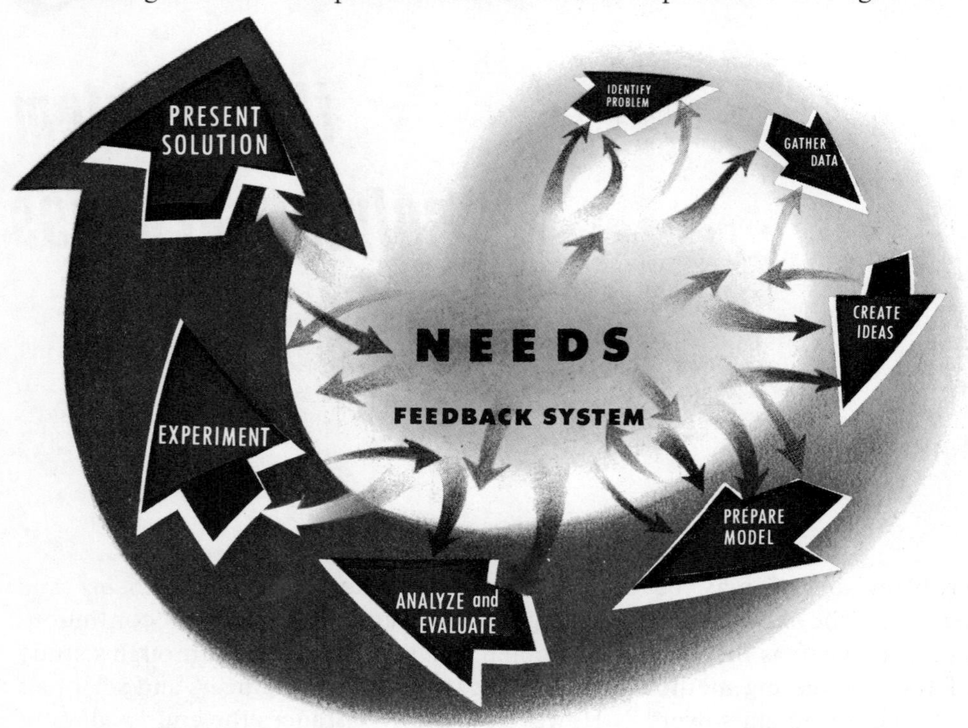

Figure 5-1. The problem solving process.

Each part of this "cyclic" process will be described in more detail, but first, two general characteristics of the process should be recognized:

1. Although the process conventionally moves in a circular direction, there is continuous "feedback" within the cycle.
2. The method of solution is a repetitious process that may be continuously refined through any desired number of cycles (Figure 5-2).

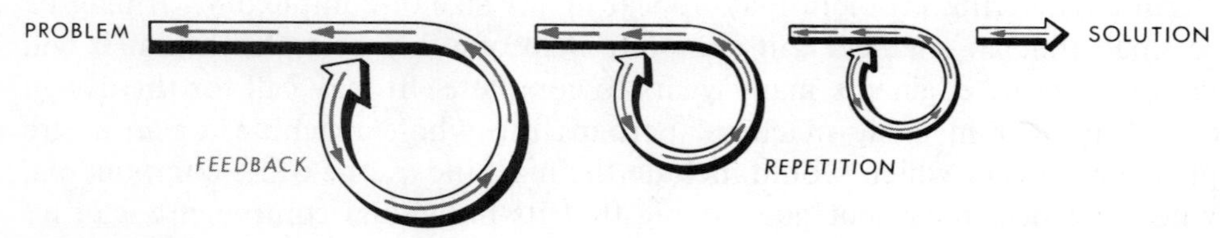

Figure 5-2

The concept of *feedback* is not new. For example, feedback is used by an individual to evaluate the results of actions that have been taken. The eye sees something bright that appears desirable and the brain sends a command to the hand and fingers to grasp it. However, if the bright object is also hot to the touch, the nerves in the fingers feed back information to the brain with the message that contact with this object will be injurious, and pain is registered to emphasize this fact. The brain reacts to this new information and sends another command to the fingers to release contact with the object. Upon completion of the feedback loop, the fingers release the object (Figure 5–3).

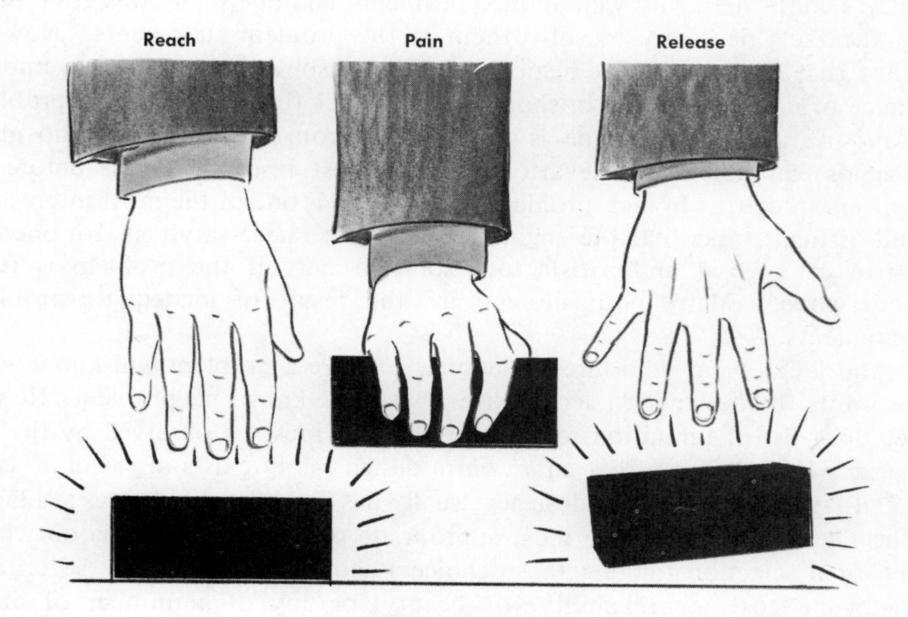

Figure 5–3

Another example is a thermostat. As part of a heating or cooling system, it is a feedback device. Changing temperature conditions produce a response from the thermostat to alter the heating or cooling rate.

The rate at which one proceeds through the problem solving cycle is a function of many factors, and these factors change with each problem. Considerable time or very little time may be spent at any point within the cycle, depending upon the situation.

Thus the problem solving process is a dynamic and constantly changing process that provides allowances for the individuality and capability of the user.

This text is concerned primarily with techniques and tools of analysis. Therefore, particular attention will be directed to steps one and four of the problem solving process–identification of the problem and preparation of a model.

IDENTIFICATION OF THE PROBLEM

One of the biggest surprises that awaits the newly graduated engineer is the discovery that there is a significant difference between the classroom problems that he solved in school and the real-life problems that he is now asked to solve. This is true because problems encountered in real life are poorly defined. The individuals who propose such problems (whether they be commercial clients or the engineer's employer) rarely know or specify exactly what is wanted, and the engineer must decide for himself what information he needs to secure in order to solve the problem. In the classroom he was confronted with well-defined problems, and he usually was given most of the facts necessary to solve them in the problem statements. Now he finds that he has available insufficient data in some areas and an overabundance of data in others. In short, he must first find out what the problem *really* is. In this sense he is no different from the physician who must diagnose an illness or the attorney who must research a case before he appears in court. In fact, problem formulation is one of the most interesting and difficult tasks that the engineer faces. It is a necessary task, for one can arrive at a good and satisfactory solution only if the problem is fully understood. Many poor designs are the result of inadequate problem statements.

The ideal client who hires a designer to solve a problem will know what he wants the designer to accomplish; that is, he knows his problem. He will set up a list of limitations or restrictions that must be observed by the designer. He will know that an *absolute* design rarely exists—a *yes* or *no* type of situation—and that the designer usually has a number of choices available. The client can specify the most appropriate optimization criteria on which the final selection (among these choices) should be based. These criteria might be cost, or reliability, or beauty, or any of a number of other desirable results.

The engineer must determine many other basic components of the problem statement for himself. He must understand not only the task that the design is required to perform but what its range of performance characteristics are, how long it is expected to last in the job, and what demands will be placed on it one year, two years, or five years in the future. He must know the kind of an environment in which the design is to operate. Does it operate continuously or intermittently? Is it subject to high temperatures, or moisture, or corrosive chemicals? Does it create noise or fumes? Does it vibrate? In short, what type of design is best suited for the job.

For example, let us assume that the engineer has been asked by a physician to design a flow meter for blood. What does he need to know before he can begin his design? Of course he should know the quantity of blood flow that will be involved. Does the physician want to measure the flow in a vein, or in an artery? Does he want to measure the flow in the very small blood vessels near the skin or in the major blood vessels leading to and from the heart? Does he want to measure the average flow of blood or the way in which the blood flow varies with every pulse beat? How easy will it be to have access to the blood vessels to be tested? Will it be better to measure the blood flow without entering the vessel itself, or should a device be inserted directly into the vessel? One major problem in inserting any kind of material into the blood stream is a strong tendency to produce blood clots. In case an instrument can be inserted into the vessel, how small must it be so that it does not disturb the flow which it is to measure? How long a section of blood vessel is available, and how does the diameter of the blood vessel vary along its length and during the measurement? These and many more components of the problem statement must be determined by the engineer before an effective solution can be designed.

Another example of the importance and difficulty of problem definition is the urban transportation problem. Designers have proposed bigger and faster subways, monorails, and other technical devices because the problem was assumed to be simply one of transporting people faster from the suburbs into the city. In many cases, it was not questioned whether the problem that they were solving was *really* the problem that needed a solution.

Surely the suburbanite needs a rapid transportation system to get into the city, but the rapid transport train is not enough. He must also have "short haul" devices to take him from the train to his home or to his work with a minimum of walking and delay. Consequently, the typical rapid transit system must be coordinated with a city-wide network of slower and shorter-distance transportation which permit the traveler to exit near his job, wherever it may be. For the suburbanite, speed is not nearly as important as frequent, convenient service, on which he can rely and for which he need not wait.

Urbanites, particularly the poor, who generally live far from the places where they might find work, are also in need of better transportation. For these people, high speed again is not nearly as important as low cost and transportation routes and vehicles that provide access to the job market. Instead of placing emphasis on bigger and faster trains, designers should consider the *wants* and the *needs* of the people they are trying to serve and determine what these wants and needs really are.

How does the engineer find out? How does he define his problem and know that his definition is in fact what is needed? Of course the first step is to find out what is already known. He must study the literature. He must become thoroughly familiar with the problem, with the environments in which it operates, with similar machines or devices built elsewhere, and with peculiarities of the situation and the operators. *He must ask questions.*

It may be, after evaluating the available information, that the engineer will be convinced the problem statement is unsatisfactory–just as today's statement of the transportation problem appears to be unsatisfactory. In that case he may suggest or perform additional studies–studies that involve the formulation of simulation models of the situation and the environment in which the machine is to be built. They may include experiments with these models to show how this environment would react to various solutions of the problem.

The design engineer must work with many types of people. Some will be knowledgeable in engineering–others will not. His design considerations will involve many areas other than engineering, particularly during problem formulation. He must learn to work with physicists and physicians, with artists, architects, and city planners, with economists and sociologists–in short, with all those who may contribute useful information to a problem. He will find that these men have a technical vocabulary different from his. They look at the world through different eyes and approach the solution of problems in a different way. It is important for the engineer to have the experience of working with such people before he accepts a position in industry, and what better opportunity is there than to make their acquaintance during his college years. With the manifold problems that tomorrow's engineer will face–problems that involve human values as well as purely technical values–collaboration between the engineer and other professional people becomes increasingly important.

PREPARATION OF A MODEL

Psychologists and others who study the workings of the human mind tell us that we can think effectively only about simple problems and small "bits" of information. They tell us that those who master complicated problems do so by reducing them to a series of simple problems which can be solved and synthesized to a final solution. This technique consists of forming a mental picture of the entire problem, and then simplifying and altering this picture until it can be taken apart into manageable components. These components must be simple and similar to concepts with which we are already familiar, to situations that we know. Such mental pictures are called *models.* They are simplified images of real things, or parts of real things–a special picture that permits us to relate it to something already known and to determine its behavior or suitability.

We are all familiar with models of sorts–with maps as models for a road system; with catalogs of merchandise as models of what is offered for sale. We have a model in our mind of the food we eat, the clothes we buy, and of the partner we want to marry.

We will often form judgments and make decisions on the basis of the model, even though the model may not be entirely appropriate. Thus the color of an apple may or may not be a sign of its ripeness, any more than the girl's apple-blossom cheeks and tip-tilted nose are the sign of a desirable girl-friend.

Engineering models are similar to sports diagrams that are composed of circles, squares, triangles, curved and straight lines, and other similar symbols which are used to represent a "play" in a football or basketball game (see Figure 5–4). Such geometrical models are limited because they are two dimensional and do not allow for the strengths, weaknesses, and imaginative decisions of the individual athletes. Their use, however, has proved to be quite valuable in simulating a brief action in the game and to suggest the best strategy for the player should he find himself in a similar situation.

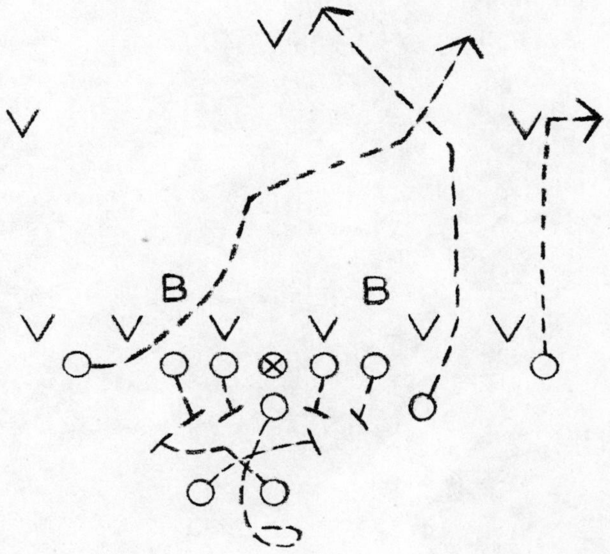

Figure 5–4

An *idealized model* may emphasize the whole of the system and minimize its component parts, or it may be designed to represent only some particular part of the system. Its function is to make visualization, analysis, and testing more practical. The engineer must recognize that he is merely limiting the complexity of the problem in order to apply known principles. Often the model may deviate considerably from the true condition; and the engineer must, of necessity, select different models to represent the same real problem. Therefore, the engineer must view his answers with respect to the initial assumptions of the model. If the assumptions were in error, or if their importance was underestimated, then the engineer's analysis will not relate closely with the true conditions. The usefulness of the model to predict future actions must be verified by the engineer. This is accomplished by experimentation and testing. Refinement and verification by experimentation are continued until an acceptable model has been obtained.

Two characteristics, more than many others, determine an engineer's competence. The first is his ability to devise simple, meaningful models; and second is the breadth of his knowledge and experience with examples with which he can compare his models. The simpler his models are, and the more generally applicable, the easier it is to predict the behavior and compute the performance of the design. *Yet models have value only to the engineer who*

Illustration 5-1. Scale models of the design are frequently tested to prove the validity of the engineer's calculations. Instrumentation on the model pictured above is given a final check by two aircraft technicians prior to testing in the Transonic Dynamics Tunnel. (Courtesy NASA [National Aeronautics and Space Administration.])

can analyze them. The beauty and simplicity of a model of the atom, Figure 5-5, will appeal particularly to someone familiar with astronomy. The free-body diagram of a wheelbarrow handle, Figure 5-6, has meaning only to someone who knows how such a diagram can be used to find the strength of the handle.

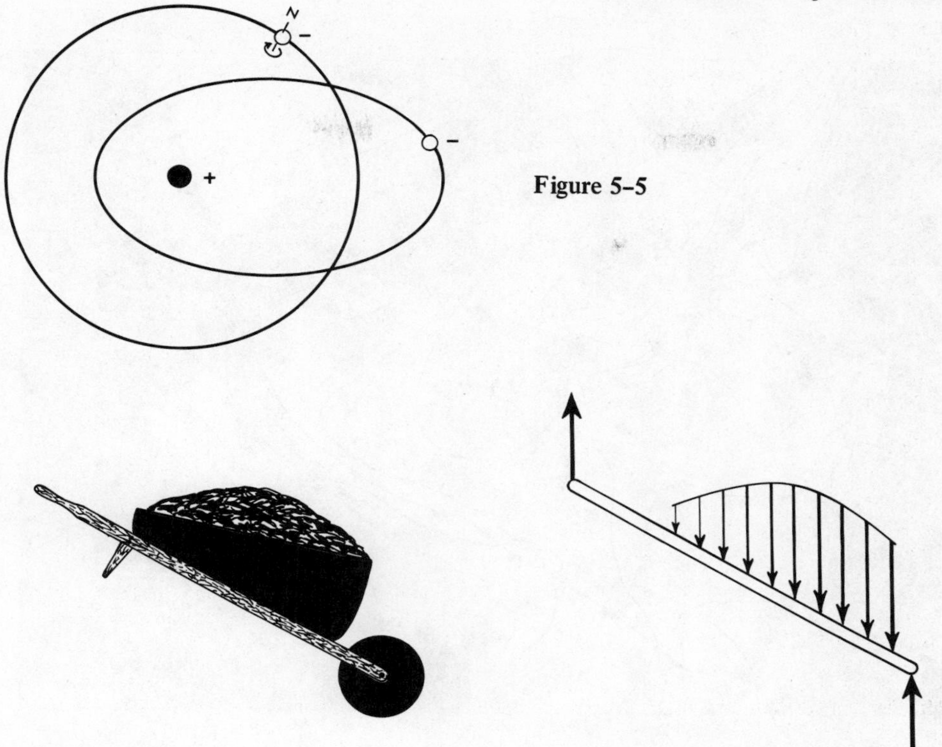

Figure 5–5

Figure 5–6

Aside from models for "things," we can made models of situations, environments, and events. The football or baseball diagram is such a model. Another familiar model of this type is the weather map, Figure 5–7, which depicts high- and low-pressure regions and other weather phenomena traveling across the country. Any meteorologist will tell you that the weather map is a very crude model for predicting weather, but that its simplicity makes the explanation of current weather trends more understandable for the layman. Models of situations and environmental conditions are particularly important in the analysis of large systems because they aid in predicting and analyzing the performance of the system before its actual implementation. Such models have been prepared for economic, military, and political situations and their preparation and testing is a science all its own.

Charts and graphs as models

Charts and graphs are convenient ways to illustrate the relationship among several variables. We have all seen charts of the fluctuations of the stock market averages, Figure 5–8, in the newspaper from day to day, or you may have had your father plot your growth on the closet door. In these examples, *time* is one of the variables. The others, in the examples above, are the average value of the stock in dollars and your height in feet and inches,

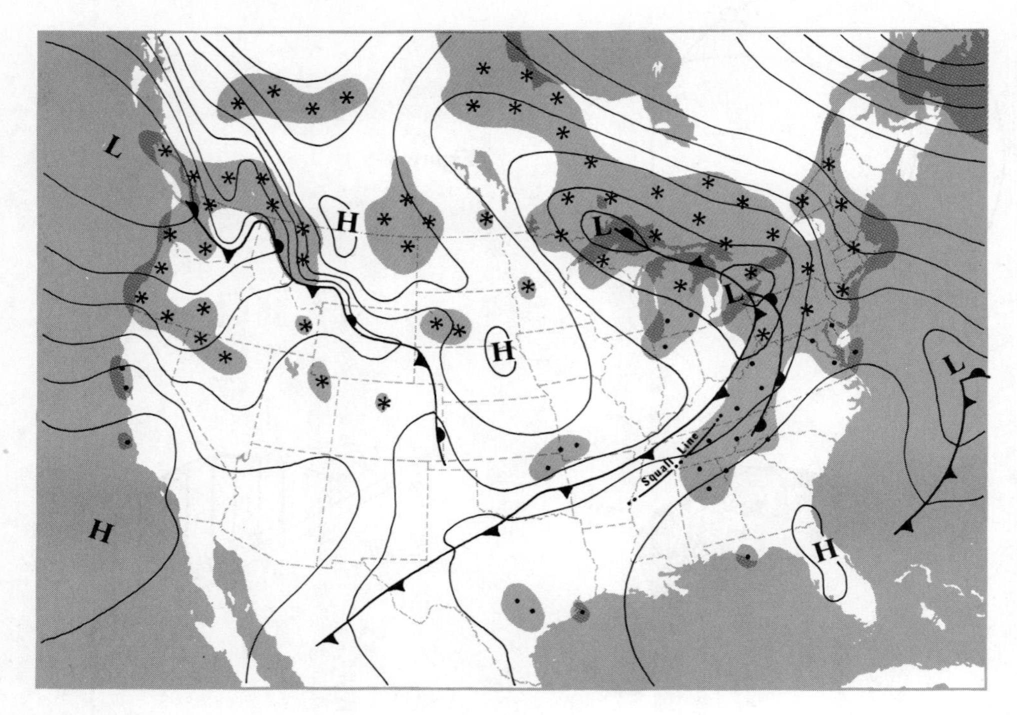

Figure 5-7

respectively. A chart or graph is not a model but presents facts in a readily understandable manner. *It becomes a model only when used to predict, project, or draw generalized conclusions* about a certain set of conditions. Consider the following example of how facts can be used to develop a chart and a graphical model. An engineer may wish to test a pump and determine

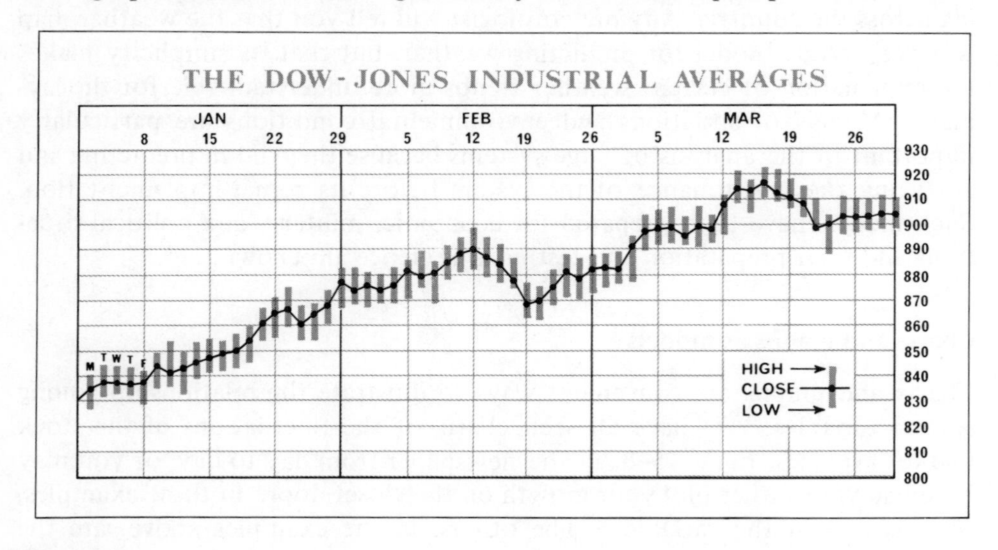

Figure 5-8

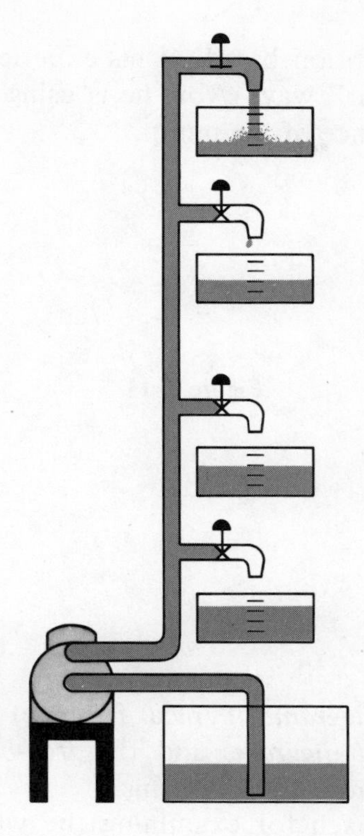

Figure 5-9

how much water it can deliver to different heights, Figure 5–9. Using a stopwatch and calibrated reservoirs at different heights, he measures the amount of water pumped to the different heights in a given time. His test results are plotted as crosses on a chart as shown in Figure 5–10. At this point, the plotted facts are a chart and not a model. Only when the engineer makes the assumption that the plotted points represent the typical performance of this or a similar pump under corresponding conditions can the chart be considered to be a graphical model. Once this assumption is made, he can draw a smooth curve through the points. With this performance curve as a model, the engineer can predict that, if he put additional reservoirs between the actual ones, they would produce results much like those shown by the

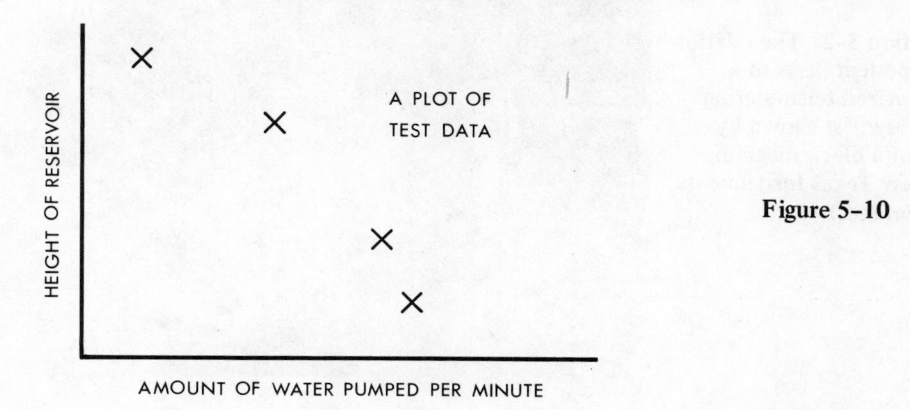

Figure 5–10

circles in Figure 5–11. He makes this assumption based on his experience that pumps are likely to behave in a "regular" way. *Now* he is using the graph as an engineering model of the performance of the pump.

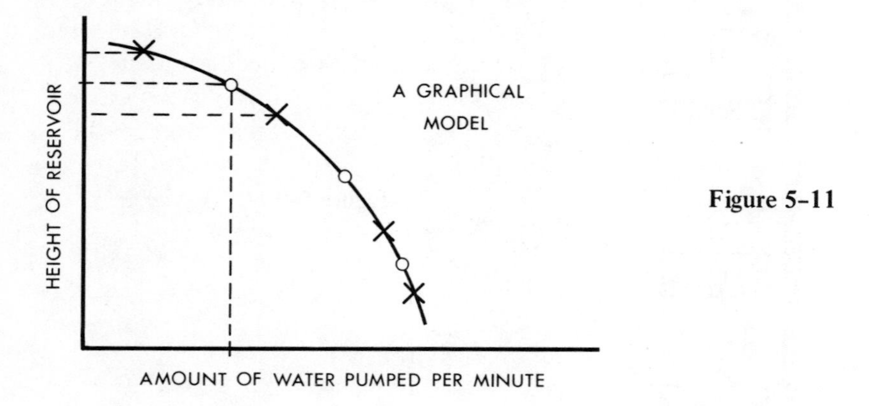

Figure 5–11

The diagram

A model often used by the engineer is the *diagram.* Typical forms of diagrams are the *block diagram,* the *electrical diagram,* and the *free-body diagram.* Some attention should be given to each of these forms.

The *block diagram* is a generalized approach for examining the whole problem, identifying its main components, and describing their relationships and interdependencies. This type of diagram is particularly useful in the

Illustration 5–2. The relation of component parts of a transistorized telemetering system are best shown by means of a block diagram. (Courtesy Texas Instruments Incorporated.)

early stages of design work when representation by mathematical equations would be difficult to accomplish. Illustration 5–2 is an example of a block diagram in which components are drawn as blocks, and the connecting lines between blocks indicate the flow of information in the whole assembly. This type of presentation is widely used to lay out large or complicated systems–particularly those involving servoelectrical and mechanical devices. No attempt is made on the drawing to detail any of the components' features. They are often referred to as "black boxes"–components whose *function* we know, but whose *details* are not yet designed.

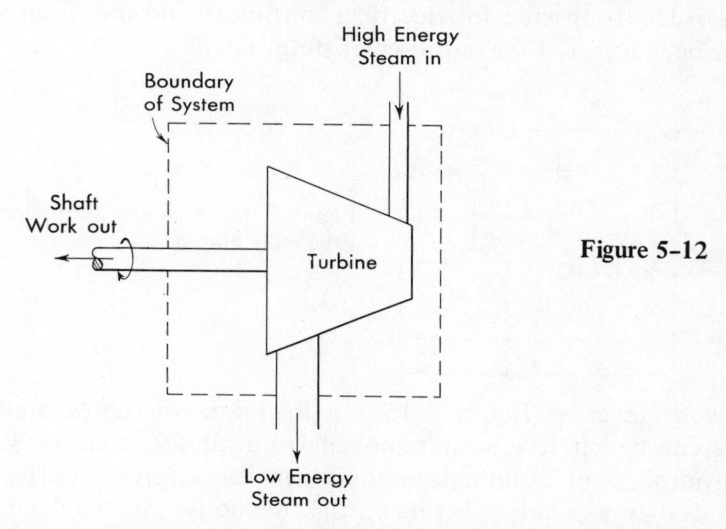

Figure 5–12

The *energy diagram* is a special form of the block diagram and is used in the study of thermodynamic systems involving mass and energy flow. Some examples of the use of an *energy diagram* are given in Figures 5–12 and 5–13.

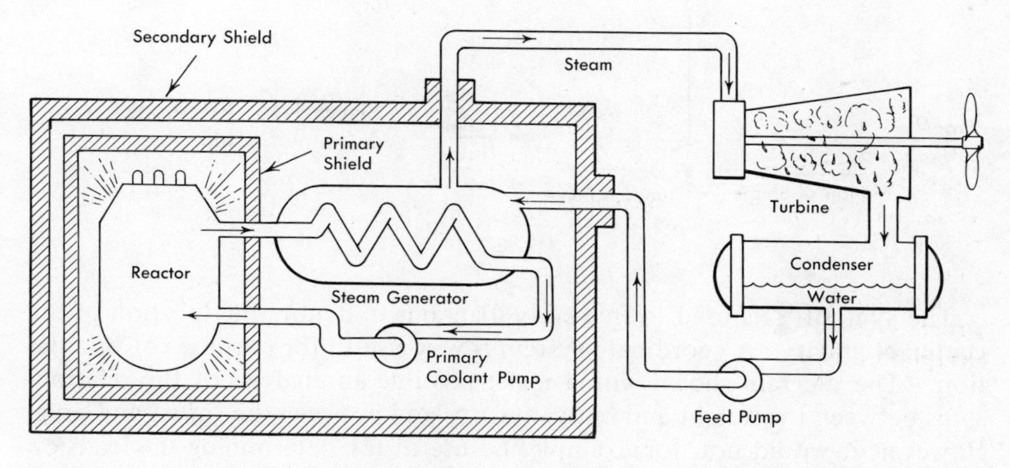

Figure 5–13. Diagramatic sketch showing how nuclear power can be used to operate a submarine. (Courtesy: General Dynamics, Electric Boat Division.)

The *electrical diagram* is a specialized type of model used in the analysis of electrical problems. This form of *idealized model* represents the existence of particular electrical circuits by utilizing conventional symbols for brevity. These diagrams may be of the most elementary type, or they may be highly complicated and require many hours of engineering time to prepare. In any case, however, they are representations or models in symbolic language of an electrical assembly.

Figure 5–14 shows an electrical diagram of a photoelectric tube that is arranged to operate a relay. Notice that the diagram details only the essential parts in order to provide for electrical continuity and thus is an idealization that has been selected for purposes of simplification.

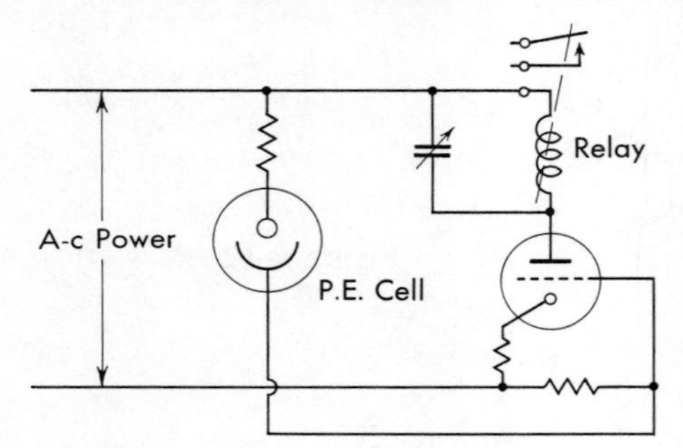

Figure 5–14. A simple photoelectric tube relay circuit.

The *free-body diagram,* Figure 5–15, is a diagrammatic representation of a physical system which has been removed from all surrounding bodies or systems for purposes of examination and where the equivalent effect of the surrounding bodies is shown as acting on the free body. Such a diagram may be drawn to represent a complex system or any smaller part of it. This form of *idealized model* is most useful in showing the effect of forces that can act upon a system. The free-body diagram will be discussed more fully below.

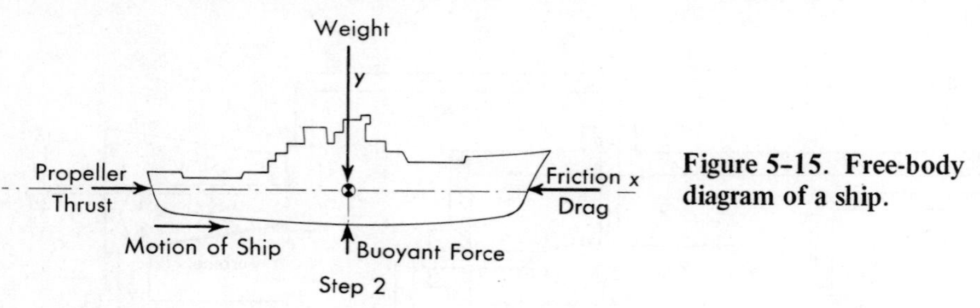

Figure 5–15. Free-body diagram of a ship.

The symbol ⊗ is used in free-body diagrams to denote the location of the center of gravity. A coordinate system is very useful for purposes of orientation. The diagram shown would make possible an analysis of the relationships between the weight and buoyant force and between the thrust and drag. However, it would not, for example, be useful for determining the loads on the ship's engine mounts. Another model (free-body diagram) of the engine alone would be required for this purpose.

General suggestions for drawing free-body diagrams

To aid the student in learning to draw free-body diagrams, the following suggestions are given:

1. **Free bodies** Be certain that the body is *free* of all surrounding objects. Draw the body so it is *free.* Do not show a supporting surface but rather show only the force vector which replaces that surface. Do not rotate the body from its original position but rather rotate the axes if necessary. Show all forces and label them. Show all needed dimensions and angles.

2. **Force components** Forces are often best shown in their component forms. When replacing a force by its components, select the most convenient directions for the components. Never show both a force and its components by solid-line vectors; use broken-line vectors for one or the other since the force *and* its components do not occur simultaneously.

3. **Weight vectors** Show the weight vector as a vertical line with its tail or point at the center of gravity, and place it so that it interferes least with the remainder of the drawing. It should always be drawn vertically.

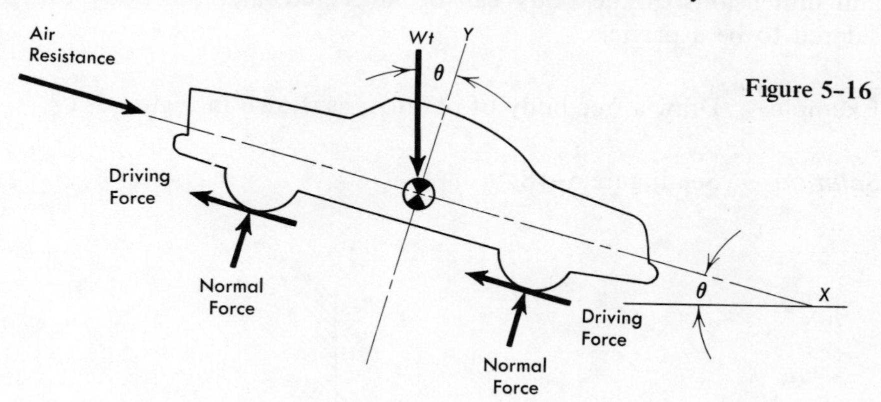

Figure 5–16

4. **Direction of vectors** The free-body diagram should represent the facts as nearly as possible. If a pull on the free body occurs, place the tail of the vector at the actual point of application and let the point of the vector be in the true direction of the pull. Likewise, if a push occurs on the free body, the vector should show the true direction, and the point of the arrow should be placed at the point of application. Force vectors on free-body diagrams are not usually drawn to scale but may be drawn proportionate to their respective magnitudes.

5. **Free-body diagram of whole structure** This should habitually be the first free body examined in the solution of any problem. Many problems cannot be solved without this first consideration. After the free body of the

whole structure or complex has been considered, select such members or subassemblies for further free-body diagrams as may lead to a direct solution.

6. Two-force members When a two-force member is in equilibrium, the forces are equal, opposite, and collinear. If the member is in compression, the vectors should point toward each other; if a member is in tension, they should point away from each other.

7. Three-force members When a member is in equilibrium and has only three forces acting on it, the three forces are always concurrent; that is, they go through the same point if they are not parallel. In analyzing a problem involving a three-force member, one should recall that any set of concurrent forces may be replaced by a resultant force. Hence, if a member in equilibrium has forces acting at three points, it is a three-force member regardless of the fact that the force applied at one or more points may be replaced by two or more components.

8. Concurrent force system For a concurrent force system the size, shape, and dimensions of the body can be neglected, and the body can be considered to be a particle.

Example Draw a free body of point A, as shown in Figure 5-17.

Solution See Figure 5-18.

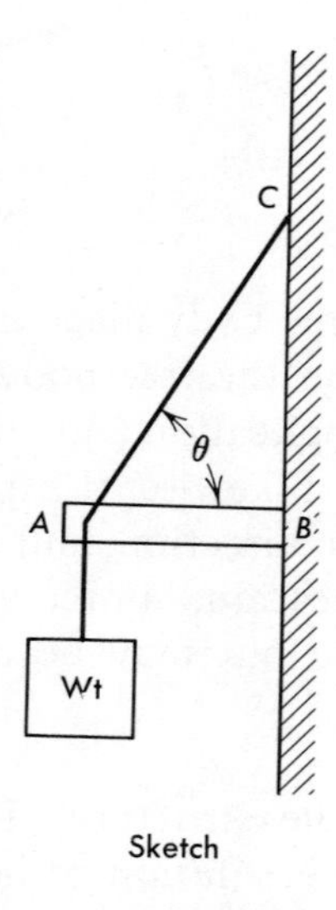

Figure 5-17

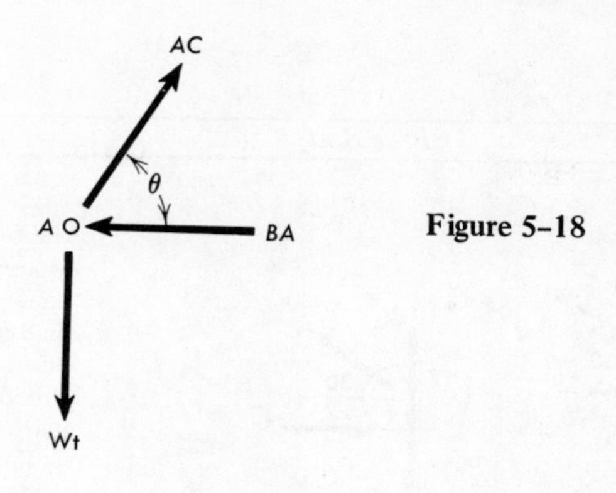

Free Body

Figure 5–18

Figure 5–19

Situation	*Free Body*	*Explanation*
A box resting on a plane Wt = 10 lb	10 lb N	The normal force always acts at an angle of 90° with the surfaces in contact. This force N usually is considered to act through the center of gravity of the body.
A weight hanging from a ring 30° 30° Wt	T_2 T_1 30° 30° Wt	Since the ring is of negligible size, it may be considered to be a point. All of the forces would act through this point. The downward force W is balanced by the tensions T_1 and T_2. The numerical sum of these tensions will be greater than the weight. This is true since T_1 is pulling against T_2.
A box on a frictionless surface Wt = 10 lb P 30°	10 lb P 30° N	Some surfaces are considered frictionless although in reality, no surface is frictionless. The force P is an unbalanced force and it will produce an acceleration. The symbol ⊗ denotes the location of the center of gravity of the body.

*See page 179 for an explanation of moments.

Figure 5-19 (cont)

Situation	*Free Body*	*Explanation*
A small box on a rough surface		The force of friction will always oppose motion or will oppose the tendency to move. For bodies of small size, the *moment effect* of the friction force may be disregarded and the friction and normal forces may be considered to act through the center of gravity of the body.
A beam resting on fixed supports		For a uniform beam, the weight acts at the midpoint of the beam regardless of where the supports are located.
A pivoted beam resting on a roller		Since a roller cannot produce a horizontal reaction, the horizontal component of any force must be counteracted by the horizontal component of the reaction at the pivoted end.
A ladder resting against a frictionless wall		At the upper end of the ladder, the only reaction possible is perpendicular to the wall since the surface is considered to be frictionless.
Pulling a barrel over a curb		All of the forces are acting through the center of the barrel.

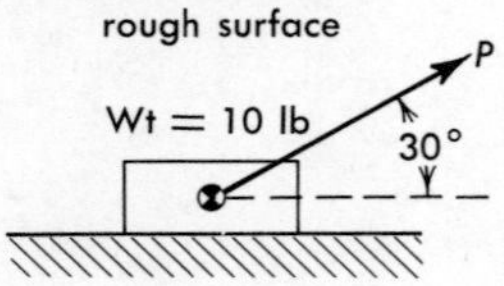

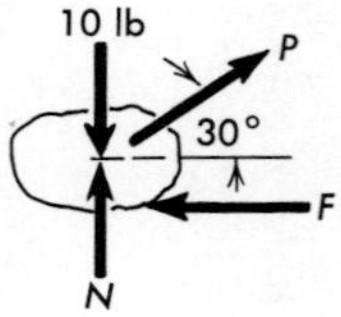

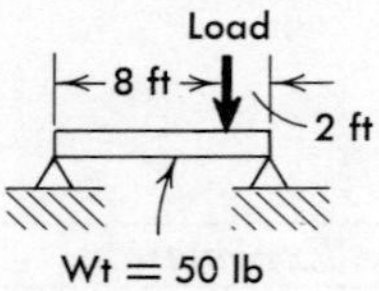

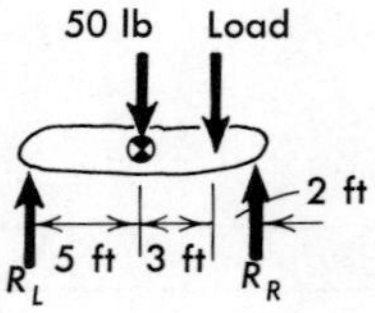

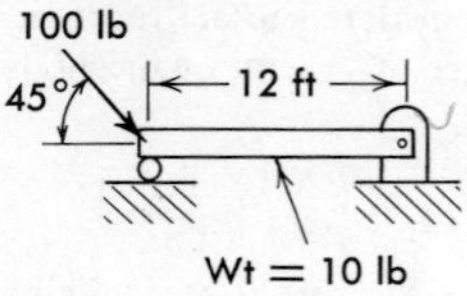

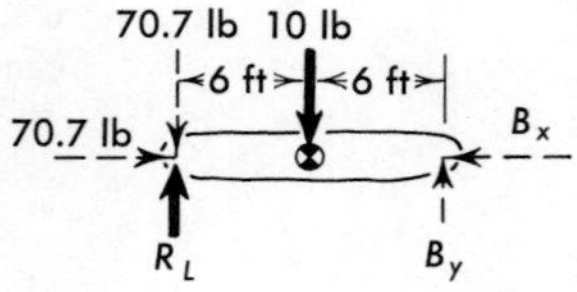

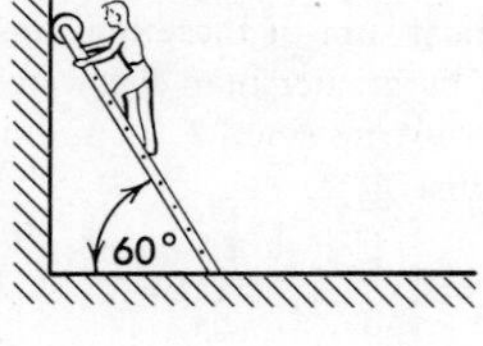

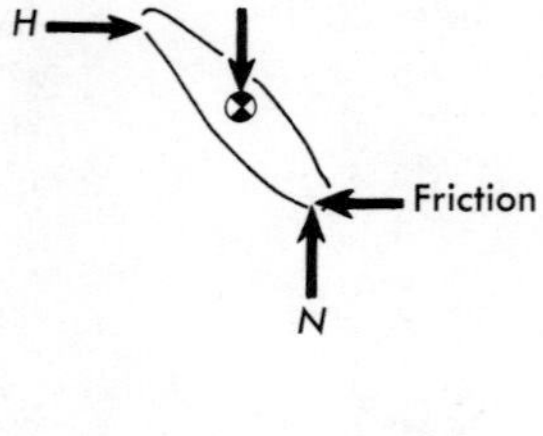

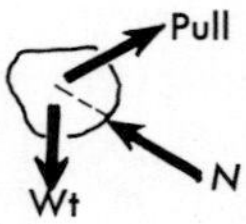

Problems

5-1. After reviewing the ecological needs of your home town, state three problems that should be solved.

5-2. Give three examples of *feedback* that existed prior to A.D. 1800.

5-3. Talk to an engineer who is working in design or development in industry. Describe two situations in his work where he has not been able to rely on *theoretical textbook solutions* to solve his problems. Why was he forced to resort to other means to solve the problems?

5-4. Describe an incident where an individual or group abandoned their course of action because it was found that they were spending time working on the wrong problem.

5-5. List the properties of a kitchen electric mixer.

5-6. List the properties of the automobile that you would like to own.

5-7. Make a matrix analysis of the possible solutions to the problem of removing dirt from clothes.

5-8. For ten minutes solo brainstorm the problem of disposal of home wastepaper. List your ideas for solution.

5-9. List five types of models that are routinely used by the average American citizen.

5-10. Diagram the model of the football play that made the longest yardage gain for your team this year.

5-11. Draw an energy system of an ordinary gas-fired hot water heater.

5-12. Draw an energy system representing a simple refrigeration cycle.

5-13. Draw an energy system representing a "perpetual motion" machine.

5-14. Draw an electrical circuit diagram containing two single-pole, double-throw switches in such manner that a single light bulb may be turned on or off at either switch location.

5-15. Arrange three single-pole, single-throw switches in an electrical circuit containing three light bulbs in such a manner that one switch will turn on one of the bulbs, another switch will turn on two of the bulbs, and the third switch will turn on all three bulbs.

5-16. Draw a free-body diagram of the container that is being lowered into the nuclear reactor, Illustration 5-3.

5-17. Draw a free-body diagram of the exercise device that has been designed to help keep astronauts in good physical shape in the weightlessness of space, Illustration 5-4.

5-18. Draw a free-body diagram of the motorcycle, Illustration 5-5.

5-19. Draw a free-body diagram of the electric motor, Illustration 5-6. What is the relationship of the force exerted against the top hook and that exerted against one of the lower hooks?

5-20. Draw a free-body diagram of the nut being tightened, Illustration 5-7.

5-21. Draw a free-body diagram of the boom that is located on the "pipe side" of the tractor, Illustration 5-8.

5-22. Draw a free-body diagram of the forces that are being exerted on the "Phillips head" screwdriver, Illustration 5-9. .

5-23. Draw a free-body diagram of the wingless, turbine-powered flying craft, Illustration 5-10.

5-24. Draw a free-body diagram of the horizontal test beam that is being tested, Illustration 5-11.

5-25. Draw a free-body diagram of the four-legged quadruped machine, Illustration 5-12. The right front leg of the unit is controlled by the operator's right arm, its left front leg by his left arm, its right rear leg by his right leg, and its left rear leg by his left leg. This research prototype is 11 ft high and weighs 3000 lb.

5-26. Describe three situations where a scale model would be the most appropriate kind of idealized model to use.

Illustration 5-3. (Courtesy Westinghouse Electric Corporation.)

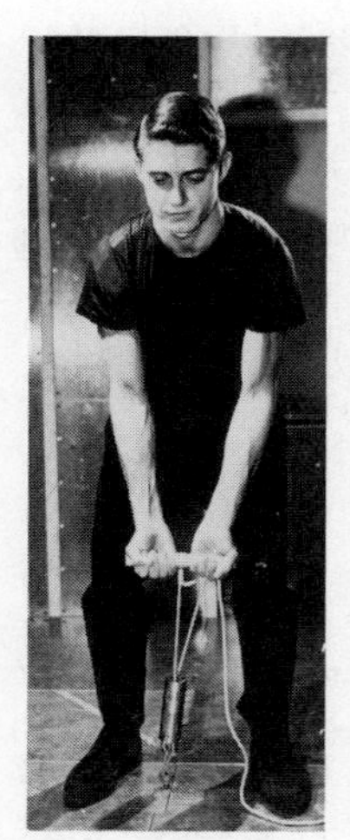

Illustration 5-4. (Courtesy Lockheed Missiles & Space Company.)

Illustration 5-5. (Courtesy *Cycle World.*)

Illustration 5-6. (Courtesy Allis Chalmers.)

Illustration 5-7. (Courtesy American Airlines.)

Illustration 5-8. (Courtesy Koppers Company, Inc.)

Illustration 5–9. (Courtesy RCA.)

Illustration 5–10. (Courtesy North American Rockwell.)

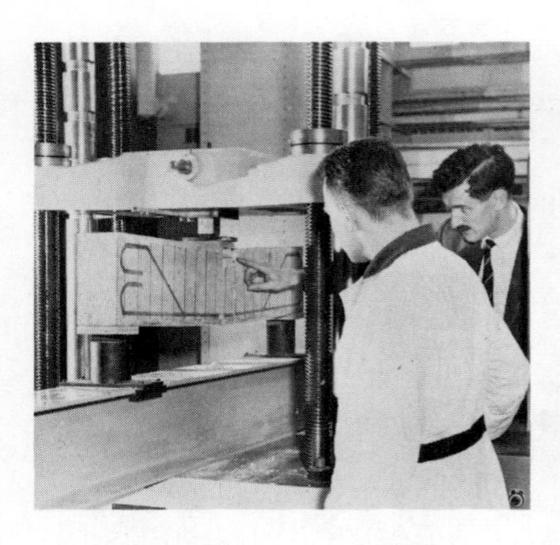

Illustration 5–11. (Courtesy Tinius Olsen Testing Machine Co., Inc.)

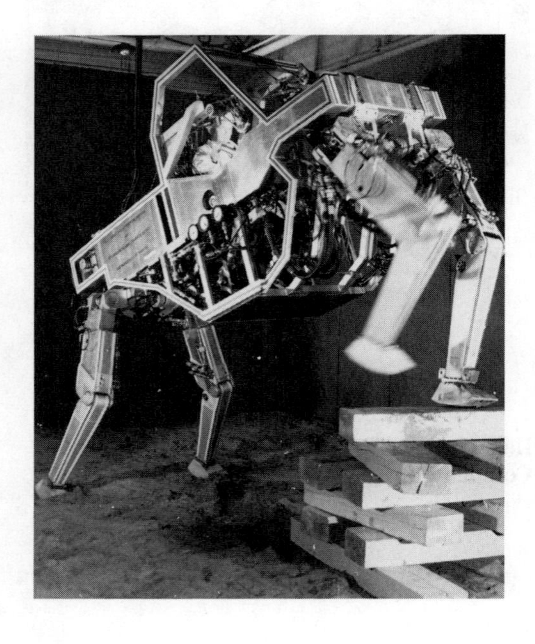

Illustration 5–12. (Courtesy General Electric Research and Development Center.)

6 problem solving

General Problems

6-1. A piece of cheese is cylindrical in shape and weighs 30 pounds. It is 0.46 metre in diameter and 12.7 cm high. If a sector is cut from it that has a 30 degree angle at the center, find the cost of the sector of cheese at 80 cents per pound. Find the number of cubic inches in the sector.

6-2. A white pine board is 14 m long and 20 cm by 80 cm in cross section. How much will the board weigh? At $160 per 1000 fbm, what is its value?

6-3. A cast iron cone used in a machine shop is 10 cm in diameter at the bottom and 34 cm high. What is the weight of the cone?

6-4. How many cubic yards of soil will it take to fill a lot 63 ft wide by 100 ft deep if it is to be raised 3 ft in the rear end and gradually sloped to the front where it is to be 1½ ft deep?

6-5. A sphere whose radius is 1.42 cm is cut out of a solid cylinder 8.8 cm high and 7.8 cm in diameter. Find the volume cut away, in cubic inches. If the ball is steel, what does it weigh?

6-6. A container is 12 in. high, 10 in. in diameter at the top, and 6 in. in diameter at the bottom. What is the volume of this container in cubic inches? What is the weight of mercury that would fill this container?

6-7. A canal on level land is 19 mi long, 22 ft deep, and has a trapezoidal cross section. The distance across the canal at the top is 36 ft and across the bottom is 15 ft. Find: (a) the number of cubic yards of dirt that were removed to complete the canal; (b) the time in hours required to pump the canal full of water if the pump discharges 600 gpm and gates at either end are closed.

6-8. A cylindrical tank 7.50 m in diameter and 15 m long is lying with its axis horizontal. Compute the weight of kerosene when it is one-third full.

6-9. A container that is in the form of a right rectangular pyramid has the following dimensions: base 26 m by 39 m, height 16 m. This container has one-half of its volume filled with ice water. Neglect the weight of the container. Find the weight of the contents.

6-10. A hemispherical container 3 m in diameter has half of its volume filled with lubricating oil. Neglecting the weight of the container, how much would the contents weigh if enough kerosene were added to fill the container to the brim?

6-11. Find the area in acres of a tract of land in the shape of a right triangle, one angle being 55°30′, and the shortest side being 1755 ft long. What length of fence will be needed to enclose the tract?

6-12. Points *A* and *B* are located on opposite corners of a building and are located so that they can be seen from point *C*. The distance *CA* is 256 m and *CB* is 312 m. The angle between lines *CA* and *CB* is 105°30′. How far apart are points *A* and *B*?

6-13. How many cubic feet of water will be needed to fill a swimming pool 40 m by 100 m. if the water is 1.5 m deep at one end and 5 m deep at the other end and the bottom slopes uniformly?

6-14. How many gallons of water will a water trough hold if it is made from an oil drum cut in half lengthwise? Dimensions are as follows: diameter of drum 1.55 feet; length of drum 3.97 feet.

6-15. A cylindrical tank is 20.8 ft in diameter, 2 m high, and is made of steel 7 mm thick. What is the area of the side and bottom of the tank? What is the weight of the tank?

6-16. A cylindrical tank is 22 feet in diameter and 8 feet high. How long will it take to fill the tank with water from a pipe which is flowing 33.3 gallons per minute?

6-17. A mixture of sand and gravel containing 6 per cent sand is desired. Two batches of mixed sand and gravel are available; one batch contains 10 per cent sand and one contains 3.5 per cent sand. How many cubic feet of each batch are required to make 100 cubic feet of a mixture containing 6 per cent sand?

6-18. A laundry tub has been constructed in the form of a frustum of a right pyramid. The bottom and top are square with the bottom being 10 inches on a side. If the tub is filled to a depth of 11 inches with water and the surface area of the water is 233 square inches, how many gallons of water are in the tub?

6-19. A conical cup has been cut from a circular sheet of paper. If the diameter of the cup is 4 inches and it is 5 inches deep, what was the area in square inches of the unused paper?

6-20. Two objects are made of the same material and have the same weights and diameters. One of the objects is a sphere 16 mm in diameter. If the other object is a right cylinder, what is its length?

6-21. A segment of a circle is cut from a sector whose central angle was 1.15 radians and whose radius was 12-7/16 inches. What is the area of the segment?

6-22. A rectangle is 4 inches wide and 8 inches long. What is the area of the smallest circle that will circumscribe the rectangle? Which is longer and by how much–the perimeter of the rectangle or the circumference of the circle?

6-23. A rectangular area 11 inches high and 13 inches long has the corners designated as A, B, C, D; A being the upper left-hand corner, B the upper right-hand corner, C the lower right-hand corner, and D the lower left-hand corner. (a) Determine the area of the circle that can be inscribed within the rectangle and the area of the circle that will circumscribe the rectangle. (b) What distance is the longer and how much: the perimeter of the circumscribed circle or the rectangle? (c) If a line were drawn from D to a point 3 inches to the right of A, what would be the length of the longest line of the triangle formed? What would be the area of each of the two areas so formed?

6-24. Convert the following Fahrenheit temperatures to Celsius temperatures. (*a*) 68°, (*b*) 98.6°, (*c*) 156°, (*d*) 359°, (*e*) 711°, (*f*) 2880°, (*g*) $4.7\ (10^4)°$, (*h*) -5°, (*i*) -255°.

6-25. Convert the following Celsius temperatures to Fahrenheit temperatures. (*a*) 20°, (*b*) 37°, (*c*) 155°, (*d*) 580°, (*e*) 8800°, (*f*) $1.22(10^5)°$, (*g*) -2°, (*h*) -40°, (*i*) -273°.

6-26. The temperature of liquid oxygen used as missile fuel is about -183°C. What is its temperature in degrees Fahrenheit?

6-27. The temperature of dry ice (solid carbon dioxide), used in shrinking metal parts to fit them together, is -78.5°C. What is the corresponding temperature in degrees Fahrenheit?

6-28. An air storage tank used in windtunnel research has a volume of 138 ft^3. How many cubic feet of air at atmospheric pressure will have to be pumped into it to raise the pressure to 185 psig?

6-29. A tight-fitting piston 3.77 in. in diameter in a closed cylinder compresses air from an initial pressure of 35 psig to 68 psig. If the final volume of the air is 14.58 $in.^3$, what will be the distance the piston moves?

6-30. A note is made for $1200 at 8 per cent interest to enable a shop to purchase a small lathe. Compute the amount of simple interest that will be paid in five years.

6-31. The proprietor of a small shop borrows $120 at 9 per cent interest for one year to pay for some tools. He repays the loan in twelve equal monthly payments on the principal and interest. What is the actual interest rate paid?

6-32. A deposit of $4500 is made in a savings account at a bank which pays 7 per cent interest compounded semiannually. (a) If no withdrawals are made, what will be the amount in the account in ten years? (b) To what rate of simple interest would this compounded interest correspond?

6-33. It is proposed to install an automatic lathe in a manufacturing plant. In order to purchase the lathe, a note for $13,500 at 10 per cent interest will have to be made. If the note is to run for ten years, what annual return must be expected from the lathe to justify the investment?

ELEMENTS OF STATIC MECHANICS

Resolutions of forces

In this initial study of static mechanics we shall deal mainly with concurrent, coplanar force systems. It is sometimes advantageous to combine two such forces into a single equivalent force, which we shall call a *resultant.* The original forces are called *components.*

Example What single force R pulling at point O will have the same effect as components F_1 and F_2? (Figure 6–1).

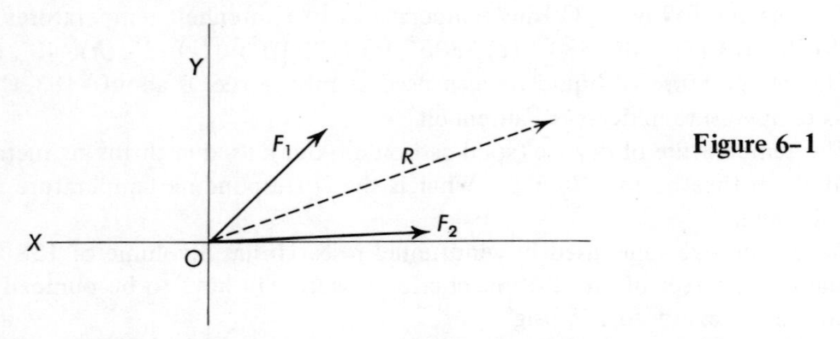

Figure 6–1

There are several methods of combining these two components into a single resultant. Let us examine one of these, the method of rectangular components.

Rectangular Components

The method most frequently used by engineers to find the resultant of force systems is the *rectangular component method.* Vector components can be added together or subtracted—always leaving some resultant value. (This resultant value, of course, may be zero.) Also, any vector or resultant value can be replaced by two or more other vectors that are usually called *components.* If the components are two in number and perpendicular to each other, they are called *rectangular components.* Although it is common practice to use space-coordinate axes that are horizontal and vertical, it is by no means necessary to do so. Any orientation of the axes will produce equivalent results.

Figure 6–2 shows a vector quantity F. Figure 6–3 shows F with its rectangular components F_x and F_y. Note that the lengths of the components F_x and F_y can be determined numerically by trigonometry. The components F_x and F_y also can be resolved into the force F by the "polygon of forces" method. Hence, they may replace the force F in any computation.

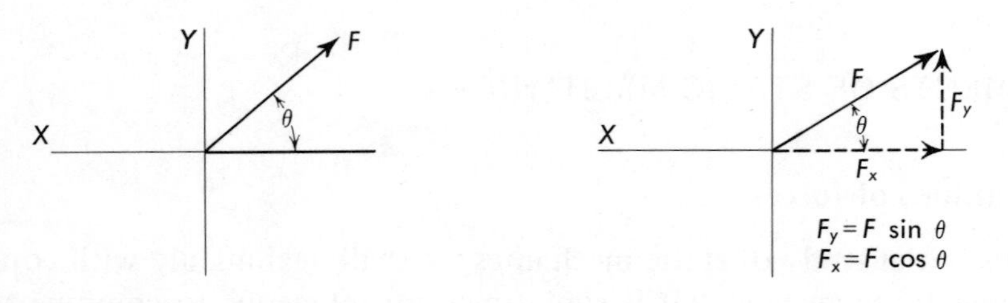

Figure 6–2 Figure 6–3

Example Let us examine a concurrent coplanar force system, Figure 6–4, and resolve each force into its rectangular components, Figure 6–5. By trigonometry, F_x can be found, using F and the cosine of the angle θ, or $F_x = F \cos \theta°$. In the same manner $F_y = F \sin \theta°$.

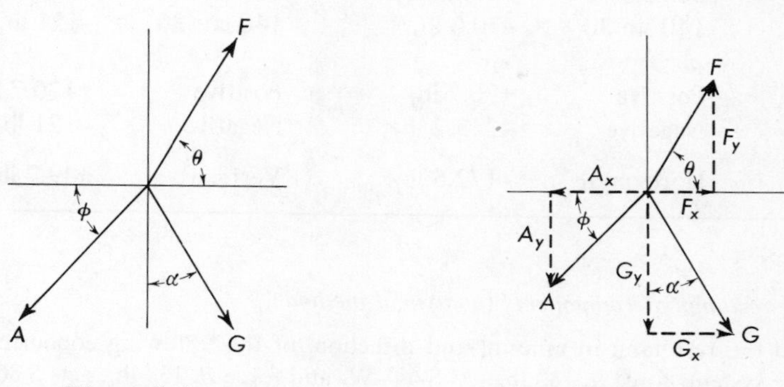

Figure 6–4 **Figure 6–5**

In order to keep the directions of the vectors better in mind, let us assume that horizontal forces acting to the right are positive and those acting to the left are negative. Also, the forces acting upward may be considered positive and those acting downward negative.

In working such force systems by solving for the rectangular components, a table may be used. When the sums of the horizontal and vertical components have been determined, lay off these values on a new pair of axes to prevent confusion. Solve for the resultant in both magnitude and direction, using the method explained on page 57.

Example Solve for the resultant, R, in Figure 6–6 using the method of rectangular components for the final resolution of the force system.

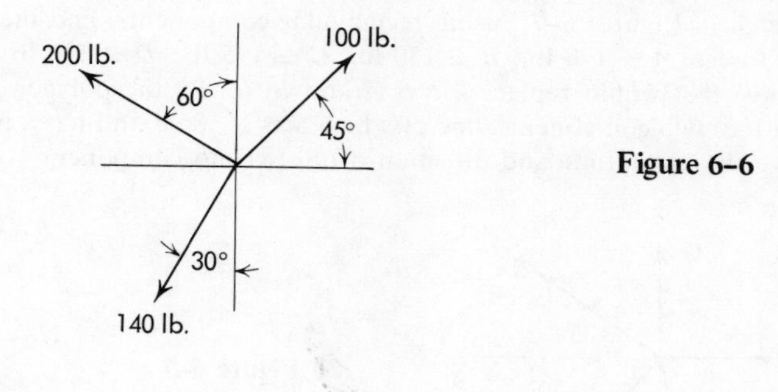

Figure 6–6

Solution

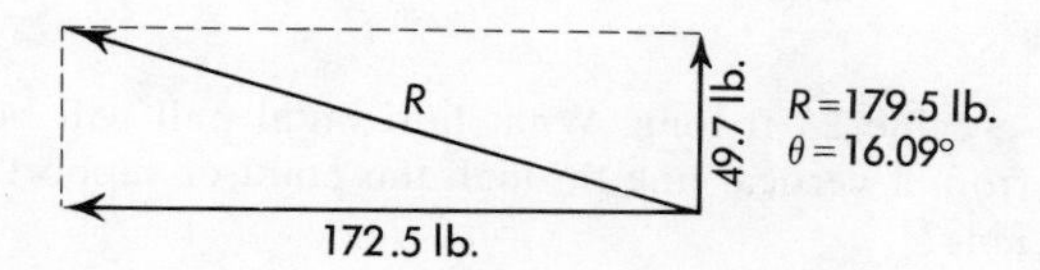

Forces	*Horizontal component*	*Horizontal value*	*Vertical component*	*Vertical value*
100 lb_f	100 cos 45° =	+70.7 lb_f	100 sin 45° =	+70.7 lb_f
200 lb_f	200 sin 60° =	-173.2 lb_f	200 cos 60° =	+100 lb_f
140 lb_f	140 sin 30° =	-70.0 lb_f	140 cos 30° =	-121 lb_f
Total value	Positive	+70.7 lb_f	Positive	+170.7 lb_f
Total value	Negative	-243.2 lb_f	Negative	-121 lb_f
Sum	Horizontal	-172.5 lb_f	Vertical	+49.7 lb_f

Problems

Solve, using rectangular components (analytical method).

6-34. Find the resultant, in amount and direction, of the following concurrent coplanar force system: force *A*, 180 lb_f acts S 60° W; and force *B*, 158 lb_f, acts S 80° W. Check graphically, using a scale of 1 in. equals 50 lb_f.

6-35. Four men are pulling a box. *A* pulls with a force of 115 lb_f, N 20°40′ E; *B* pulls with a force of 95 lb_f S 64°35′ E; *C* pulls with a force of 140 lb_f N 40°20′ E; and *D* pulls with a force of 68 lb_f E. In what direction will the box tend to move?

6-36. Determine the amount and direction of the resultant of the concurrent coplanar force system as follows: force *A*, 10 lb_f, acting N 55° E; force *B*, 16 lb_f, acting due east; force *C*, 12 lb_f, acting S 22° W; force *D*, 15 lb_f, acting due west; force *E*, 17 lb_f, acting N 10° W.

6-37. Find the resultant and the angle the resultant makes with the vertical, using the following data: 10 lb_f, N 18° W; 5 lb_f, N 75° E; 3 lb_f, S 64° E; 7 lb_f, S 0° W; 10 lb_f, S 50° W.

6-38. Five forces act on an object. The forces are as follows: 130 lb_f, 0°; 170 lb_f, 90°; 70 lb_f, 180°; 20 lb_f, 270°; 300 lb_f, 150°. The angles are measured counterclockwise with reference to the horizontal through the origin. Determine graphically the amount and direction of the resultant by means of the polygon of forces. Check analytically, using horizontal and vertical components. Calculate the angle that *R* makes with the horizontal.

6-39. (*a*) In the sketch in Figure 6-7, using rectangular components, find the resultant of these four forces: $A = 100$ lb_f, $B = 130$ lb_f, $C = 195$ lb_f, $D = 138$ lb_f. (*b*) Find a resultant force that would replace forces *A* and *B*. (*c*) By the polygon of forces, break force *A* into two components, one of which acts N 10° E and has a magnitude of 65 lb_f. Give the magnitude and direction of the second component.

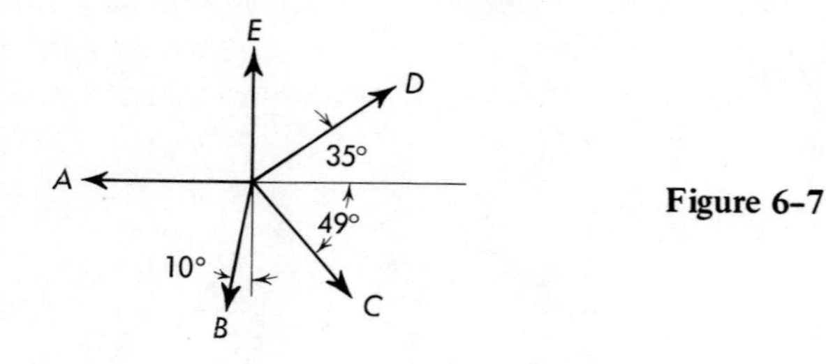

Figure 6-7

6-40. A weight of 1200 lb_f is hung by a cable 23 ft long. What horizontal pull will be necessary to hold the weight 8 ft from a vertical line through the point of support? What will be the tension in the cable?

MOMENTS

If a force is applied perpendicular to a pivoted beam some distance away from the pivot point there will be a tendency to cause the beam to turn in either a clockwise or counterclockwise direction (see Figure 6–8). The direction of the tendency will depend on the direction of the applied force. This tendency of a force to cause rotation about a given center is called *moment* (see Figure 6–9).

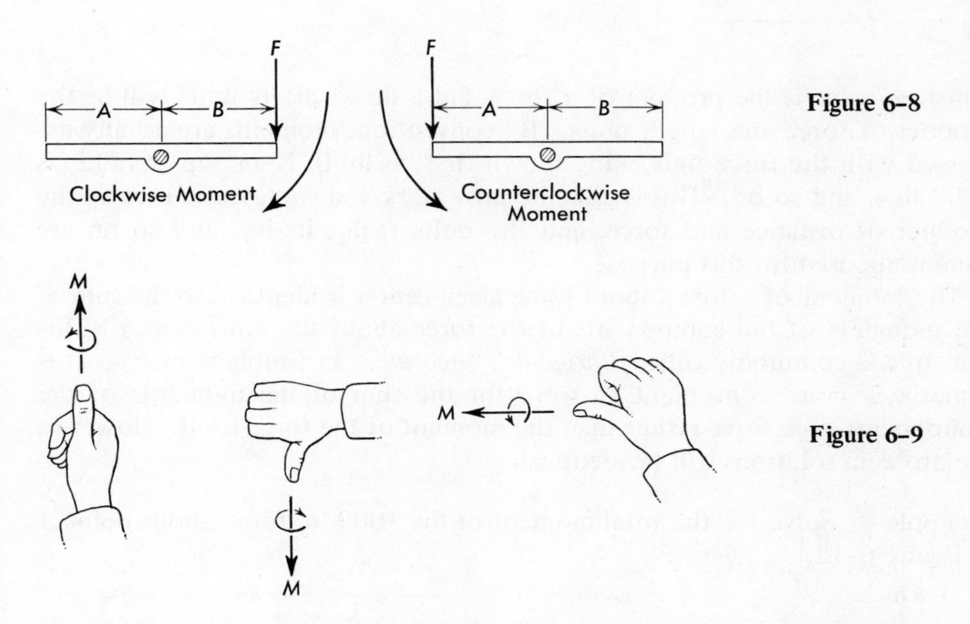

Figure 6–8

Figure 6–9

The amount of *moment* will depend upon the magnitude of the applied force as well as upon the length of the moment arm. The moment arm is the perpendicular distance from the point of rotation to the applied force. The magnitude of the moment is calculated by multiplying the force by the moment arm.

The sign convention being used in a given problem analysis should be placed on the calculation sheet adjacent to the problem sketch. In this way no confusion will arise in the mind of the reader concerning the sign convention being used. We shall assume that vectors acting to the right have a positive sense, the sketch shown in Figure 6–10 will serve as a basis for problem analysis in this text.

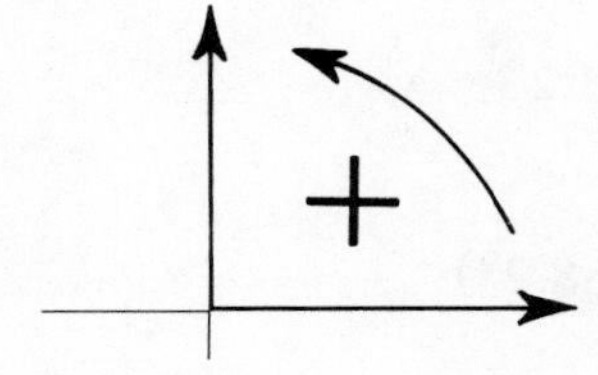

Figure 6–10

Example Solve for the moments in Figure 6–11 that tend to cause turning of the beam about the axle.

$$\text{Counterclockwise moment} = (\ 50 \text{ lb})(2 \text{ ft}) = \mathbf{+\,100 \text{ lb-ft}}$$
$$\text{Clockwise moment} = (100 \text{ lb})(5 \text{ ft}) = \mathbf{-\,500 \text{ lb-ft}}$$

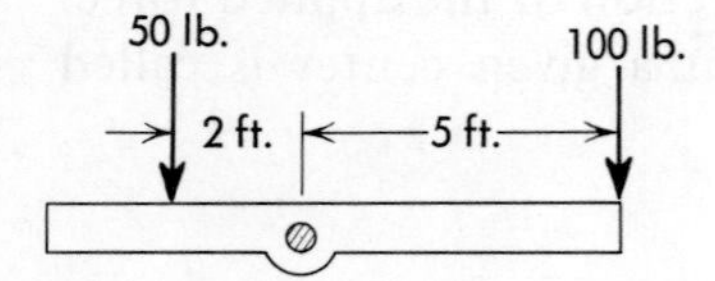

Figure 6–11

Since *moment* is the product of a force and a distance, its units will be the product of force and length units. By convention, moments are usually expressed with the force unit being shown first, as lb_f-ft, N–m, kip–in (a kip is 1000 lb_f), and so on. This is done because *work* and *energy* also involve the product of distance and force, and the units ft-lb_f, in.-lb_f, and so on are commonly used for this purpose.

The moment of a force about some given center is identical to the sum of the moments of the components of the force about the same center. This principle is commonly called *Varignon's theorem.* In problem analysis it is sometimes more convenient to solve for the sum of the moments of the components of a force rather than the moment of the force itself. However, the problem solutions will be identical.

Example Solve for the total moment of the 1000-lb_f force about point A in Figure 6–12.

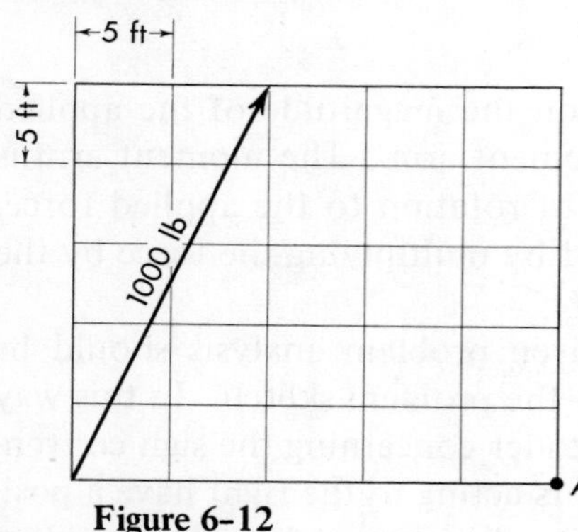

Figure 6–12

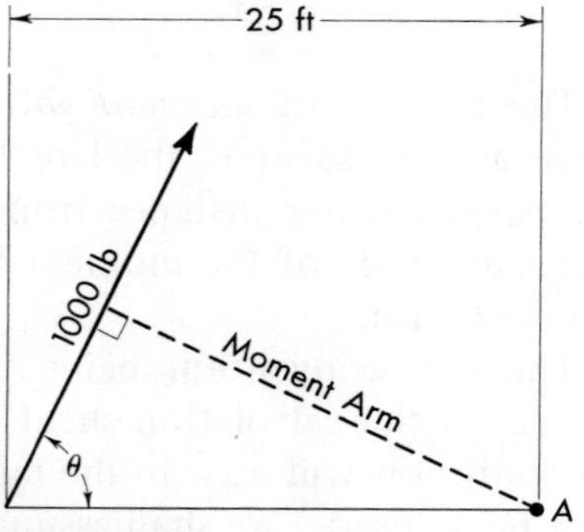

Figure 6–13

Solution A Moment of a force as shown in Figure 6–13.

$$\theta = \text{arc tan } \frac{25}{10} = 68.2°$$

$$\text{Moment arm} = 25 \sin 68.2°$$
$$\text{Total moment} = (1000)(25 \sin 68.2°)$$
$$= \mathbf{23{,}200 \text{ lb}_f\text{-ft}}$$

Solution B Moments of components of a force as shown in Figure 6–14.

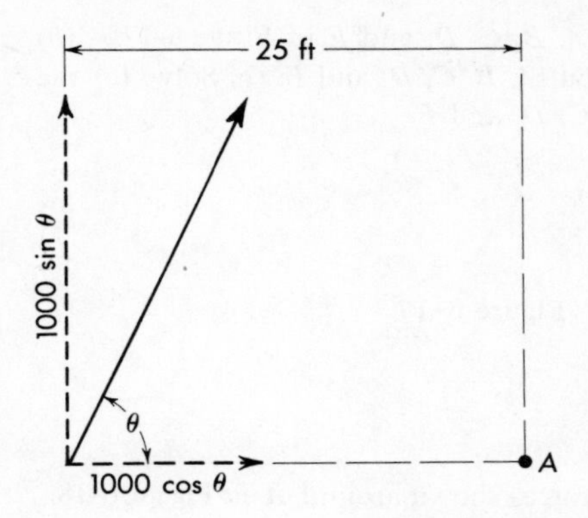

Figure 6–14

$$\text{Vertical component} = 1000 \sin 68.2°$$

and

$$\text{Moment arm} = 25 \text{ ft}$$

$$\text{Horizontal component} = 1000 \cos 68.2°$$

and

$$\text{Moment arm} = 0$$

(Note that the horizontal component passes through the center A.)

$$\text{Total moment} = (1000 \sin 68.2°)(25) = \mathbf{23{,}200\ lb_f\text{-}ft}$$

Problems

6–41. Solve for the algebraic sum of the moments in pound-feet about A when h is 20 in. as shown in Figure 6–15.

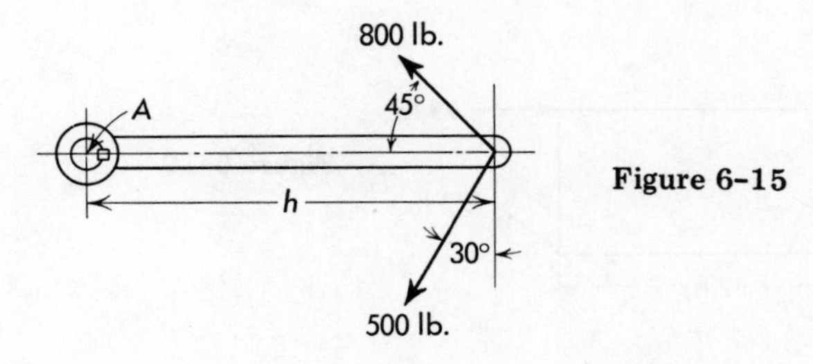

Figure 6–15

6–42. Solve for the algebraic sum of the moments about the center of the axle shown in Figure 6–16.

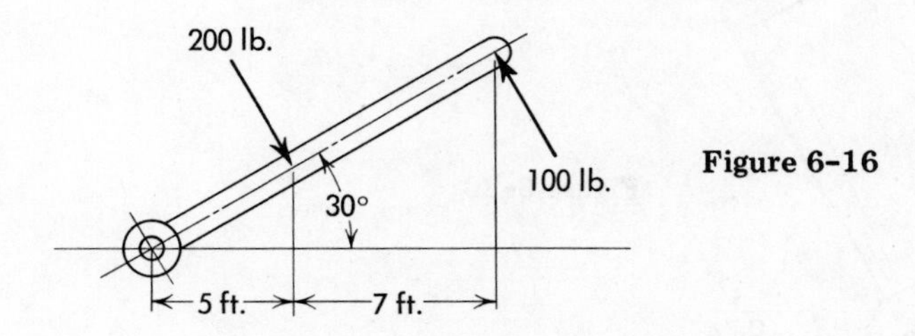

Figure 6–16

6-43. (*a*) Solve for the clockwise moments about *A*, *B*, *C*, *D*, and *E* in Figure 6-17. (*b*) Solve for the counterclockwise moments about *A*, *B*, *C*, *D*, and *E*. (*c*) Solve for the algebraic sum of the moments about *A*, *B*, *C*, *D*, and *E*.

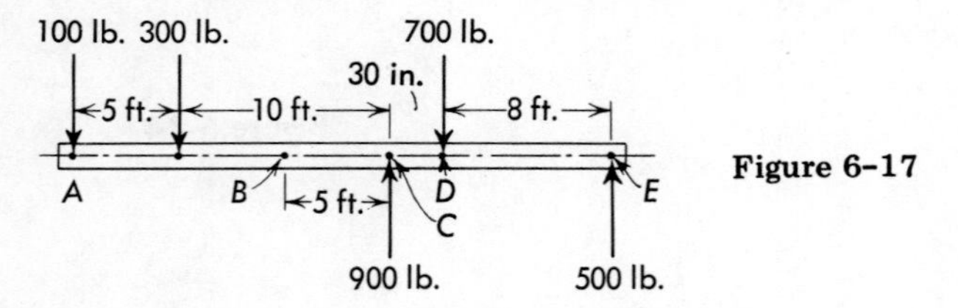

Figure 6-17

6-44. Find the summation of the moments of the forces shown around *A* in Figure 6-18. Find the moment sum around *D*.

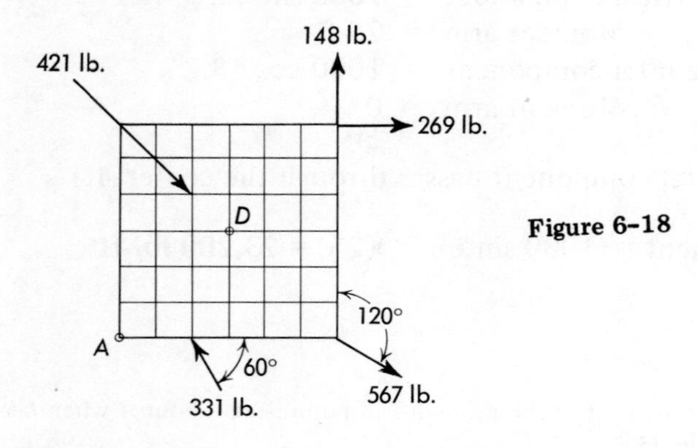

Figure 6-18

6-45. Find the moment of each of the forces shown about *O* in Figure 6-19.

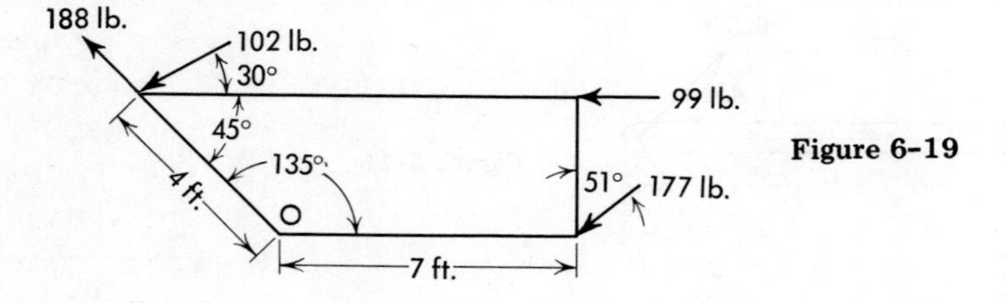

Figure 6-19

6-46. What pull *P* is required on the handle of a claw hammer to exert a vertical force of 750 lb_f on a nail? Dimensions are shown on Figure 6-20.

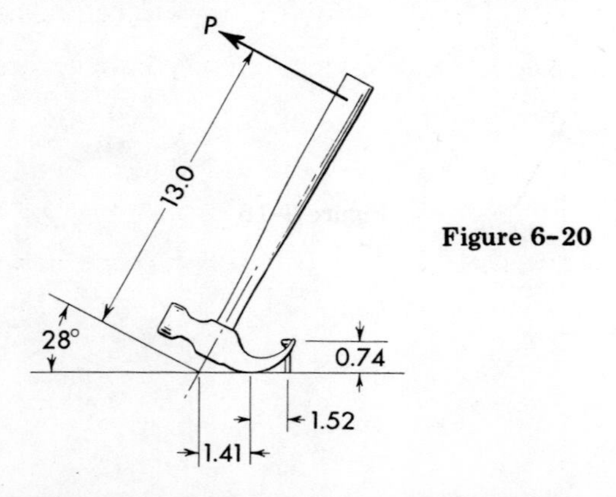

Figure 6-20

6-47. On the trapezoidal body shown in Figure 6-21 find the moment of each of the forces about point O.

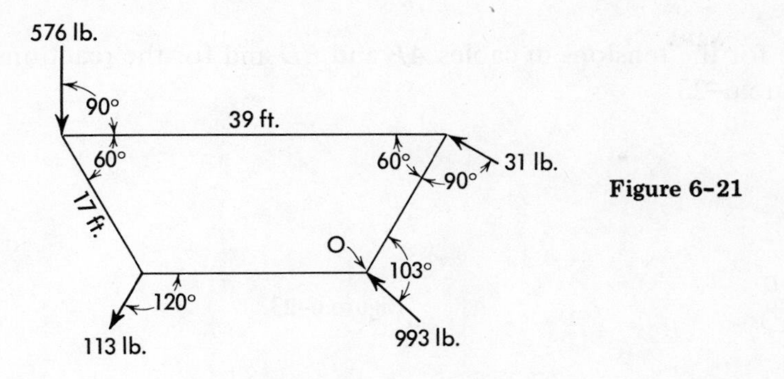

Figure 6-21

6-48. Find the moment of each of the forces shown in Figure 6-22 about the point A.

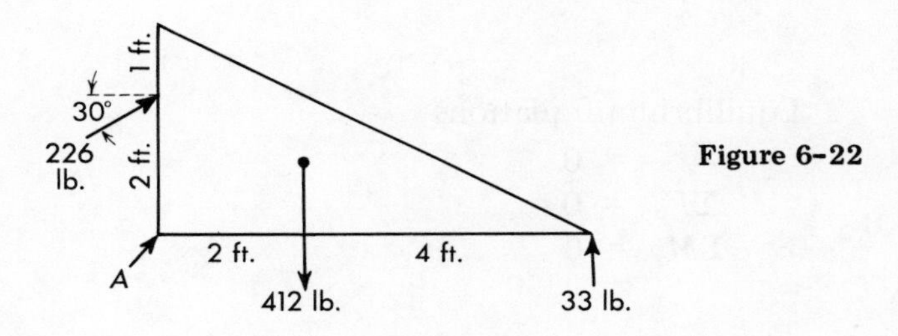

Figure 6-22

EQUILIBRIUM

The term *equilibrium* is used to describe the condition of any body when the resultant of all forces acting on the body equals zero. For example, the forces acting upward on a body in equilibrium must be balanced by other forces acting downward on the body. Also, the forces acting horizontally to the right are counteracted by equal forces acting horizontally to the left. Since no unbalance in moment or turning effect can be present when a body must also be zero. The moment center may be located at any convenient place on the body or at any place in space. We may sum up these conditions of equilibrium by the following equations.[1]

$\Sigma F_x = 0$ (the sum of all horizontal forces acting on the body equals zero)

$\Sigma F_x = 0$ (the sum of all vertical forces acting on the body equals zero)

$\Sigma M_o = 0$ (the sum of the moments of all forces acting on the body equals zero)

[1] These equations are applicable for two-dimensional problems—or force systems that lie in the plane of this paper.

These equilibrium equations may be used to good advantage in working problems involving beams, trusses, and levers.

Example Solve for the tensions in cables AF and ED and for the reactions at C and R in Figure 6–23.

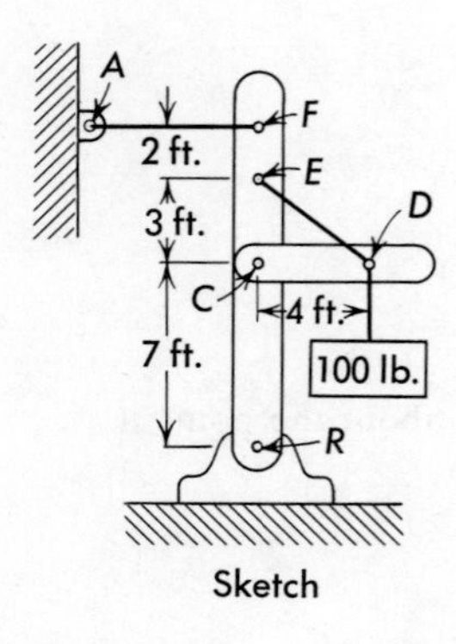

Figure 6–23

Equilibrium Equations

$$\Sigma F_x = 0$$
$$\Sigma F_y = 0$$
$$\Sigma M_o = 0$$

Solution

1. Take moments about point R in free body No. 1, Figure 6–24.

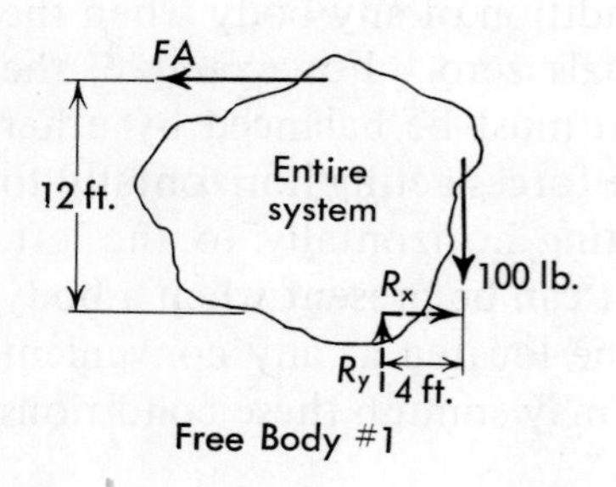

Figure 6–24

$$\Sigma M_r = 0$$
$$(12 \text{ ft})(FA) - (100 \text{ lb}_f)(4 \text{ ft}) = 0$$
$$FA = \frac{400 \text{ lb}_f\text{-ft}}{12 \text{ ft}} = 33.3 \text{ lb}_f$$

$$\Sigma F_x = 0$$
$$R_x - FA = 0$$
$$R_x = FA = \mathbf{33.3 \text{ lb}_f}\rightarrow$$

2. Take moments about point C in free body No. 2, Figure 6–25.

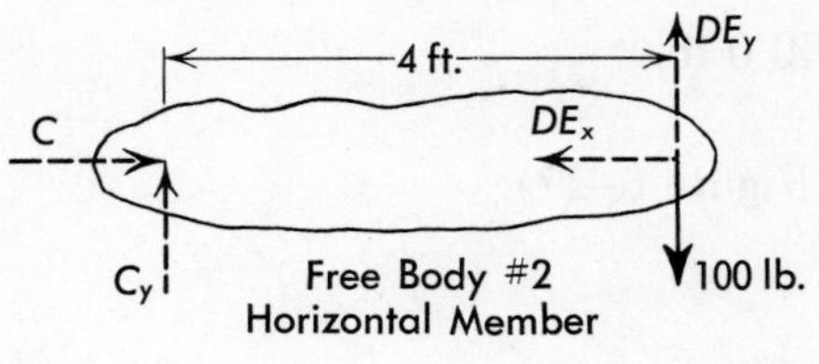

Figure 6–25

$$\Sigma M_c = DE_y\,(4) - 100\,(4) = 0$$

$$DE_y = 100\ \text{lb}_f$$

Therefore

$$DE = \frac{100\ \text{lb}_f}{\sin 36.9^\circ} = \mathbf{166.8\ lb_f} \searrow$$

And free body No. 2

$$\Sigma F_y = 0$$
$$C_y = 100\ \text{lb}_f - 100\ \text{lb}_f$$
$$= \mathbf{0}$$

Also free body No. 2

$$\Sigma F_x = 0$$

$$C_x = DE_x = \frac{100\ \text{lb}_f}{\tan 36.9^\circ}$$

$$= 133.1\ \text{lb}_f \rightarrow$$

3. Consider $\Sigma F_y = 0$, using the third free body (vertical member) as shown in Figure 6–26. Remember that in two-force members, such as cable DE, the reactions at each end will be equal in magnitude but opposite in direction; that is, E_x and E_y are equal to DE_x and DE_y.

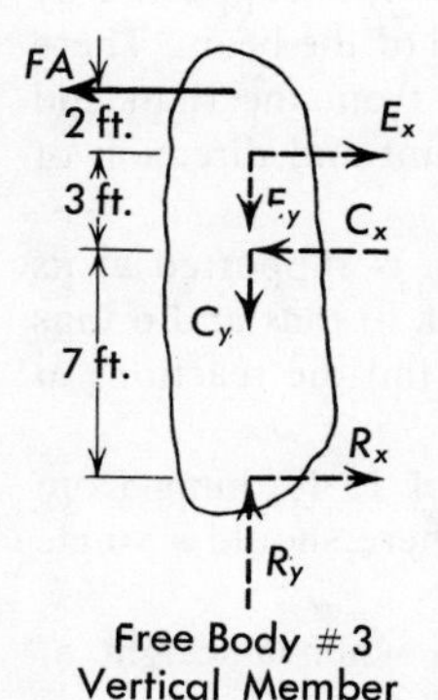

Figure 6–26

$$\Sigma F_y = 0$$
$$R_y - DE_y = 0$$
$$R_y = \mathbf{100.0\ lb_f} \uparrow$$

The resultant is indicated as before (see Figure 6–27).

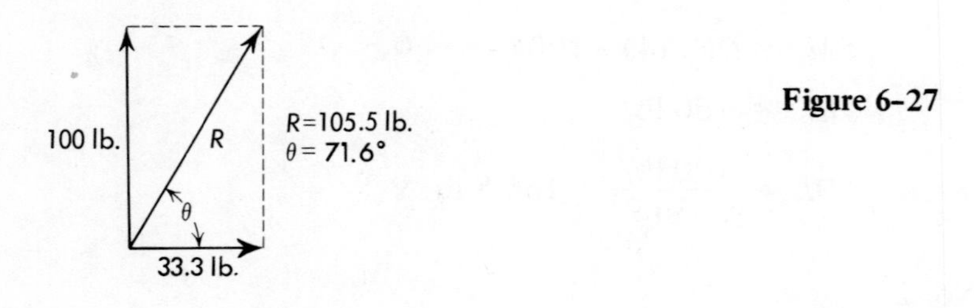

Figure 6–27

Problems

6–49. A horizontal beam 20 ft long weighs 150 lb_f. It is supported at the left end and 4 ft from the right end. It has the following concentrated loads: at the left end, 200 lb_f; 8 ft from the left end, 300 lb_f; at the right end, 400 lb_f. Calculate the reactions at the supports.

6–50. A horizontal beam 8 ft long and weighing 30 lb_f is supported at the left end and 2 ft from the right end. It has the following loads: at the left end, 18 lb_f; 3 ft from the left end, 22 lb_f; at the right end, 15 lb_f. Compute the reactions at the supports.

6–51. A beam 22 ft long weighing 300 lb_f is supporting loads of 700 lb_f 3 ft from the left end and 250 lb_f 7 ft from the right end. One support is at the left end. How far from the right end should the right support be placed so that the reactions at the two supports will be equal?

6–52. A beam 18 ft long is supported at the right end and at a point 5 ft from the left end. It is loaded with a concentrated load of 250 lb_f located 2 ft from the right end and a concentrated load of 450 lb_f located 9 ft from the right end. In addition, it has a uniform load of 20 lb_f per linear foot for its entire length. Find the reactions at the supports.

6–53. A 12-ft beam which weighs 10 lb_f per foot is resting horizontally. The left end of the beam is pinned to a vertical wall. The right end of the beam is supported by a cable that is attached to the vertical wall 6 ft above the left end of the beam. There is a 200-lb_f concentrated load acting vertically downward 3 ft from the right end of the beam. Determine the tension in the cable and the amount and direction of the reaction at the left end of the beam.

6–54. A steel I-beam, weighing 75 lb_f per linear foot and 20 ft long, is supported at its left end and at a point 4 ft from its right end. It carries loads of 10 tons and 6 tons at distances of 5 ft and 17 ft, respectively, from the left end. Find the reactions at the supports.

6–55. A horizontal rod 8 ft long and weighing 12 lb_f has a weight of 15 lb_f hung from the right end, and a weight of 4 lb_f hung from the left end. Where should a single support be located so the rod will balance?

6–56. A uniform board 22 ft long will balance 4.2 ft from one end when a weight of 61 lb_f is hung from this end. How much does the board weigh?

6-57. An iron beam 12.7 ft long weighing 855 lb_f has a load of 229 lb_f at the right end. A support is located 7.2 ft from the load end. (*a*) How much force is required at the opposite end to balance it? (*b*) Disregarding the balancing force, calculate the reactions on the supports if one support is located 7 ft from the left end and the other support is located 4 ft from the right end.

6-58. A horizontal rod 8 ft long and weighing 1.2 lb_f per linear foot has a weight of 15 lb_f hung from the right end, and a weight of 4 lb_f hung from the left end. Where should a single support be located so the rod will balance?

6-59. A 2-ft diameter sphere weighs 56 lb_f, is suspended by a cable, and rests against a vertical wall. If the cable *AB* is 2 ft long, (*a*) calculate the angle the cable will make with the smooth wall, (*b*) solve for the tension in the cable and the reaction at *C* in Figure 6–28. Check results graphically.

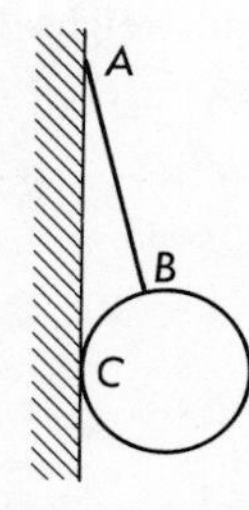

Figure 6–28

6-60. What horizontal pull *P* will be necessary just to start the wheel weighing 1400 kg_f over the 4-in. block in Figure 6–29?

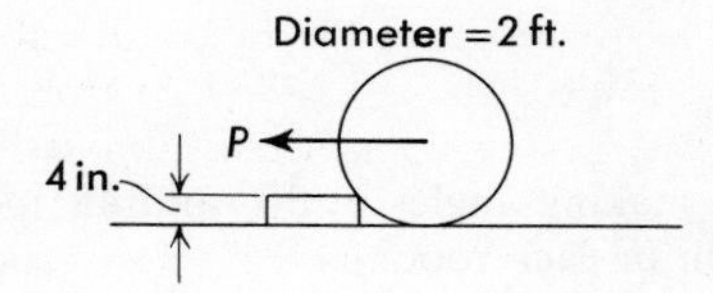

Figure 6–29

6-61 If the tension in the cable *AB* in Figure 6–30 is 196 kg_f, how much does the sphere *B* weigh? How much is the reaction of the inclined plane on the sphere?

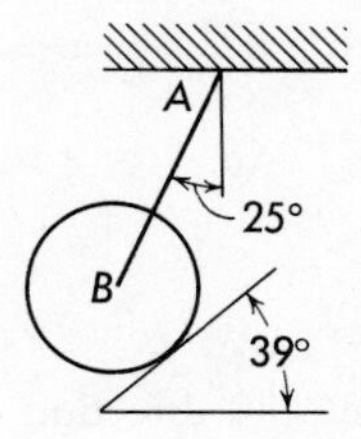

Figure 6–30

6-62. A cylinder weighing 206 kg_f is placed in a smooth trough as shown in Figure 6–31. Find the two supporting forces.

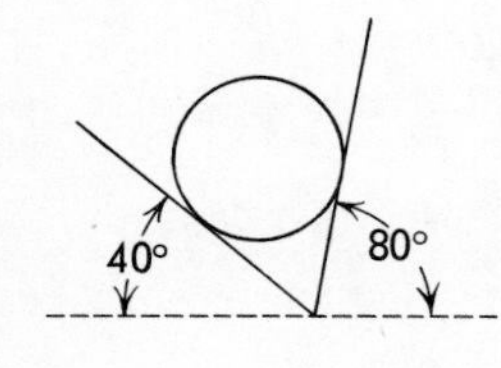

Figure 6–31

6-63. A 7960 N load is supported as shown in Figure 6-32. AB equals 3 m, θ equals 25°. (*a*) Neglecting the weight of the beam AB, solve analytically for the tension in the cable and the reaction at A. (*b*) If beam AB is uniform and weighs 15 N per m, solve for the tension in the cable and the reaction at A.

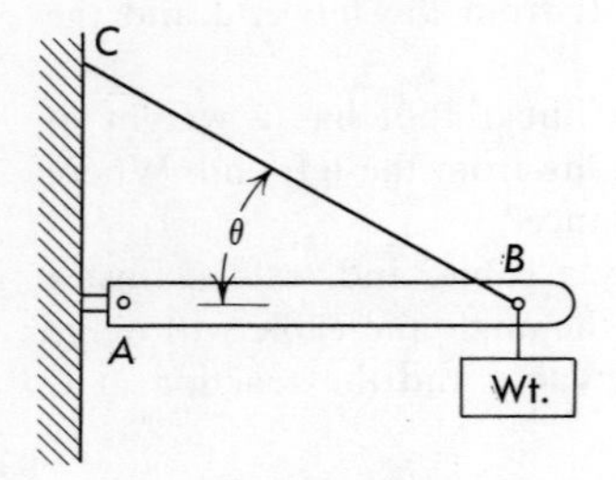

Figure 6-32

6-64. Find the tension in AB and the compression in BC in Figure 6-33.

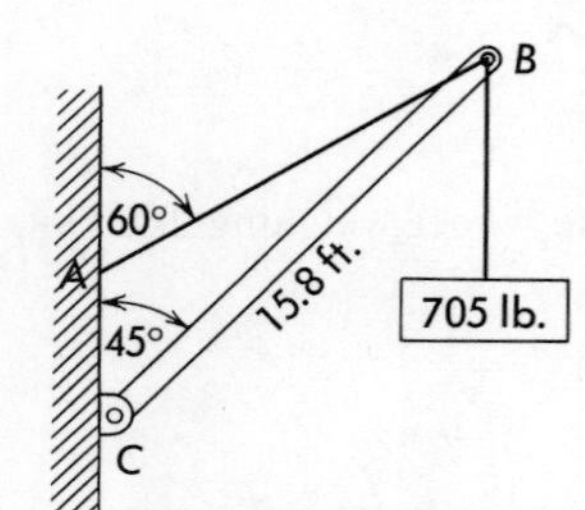

Figure 6-33

6-65. A weight of 4355 N is supported by two ropes making angles of 30° and 45° on opposite sides of the vertical. What is the tension in each rope?

6-66. Forces are applied on a rigid frame as shown in Figure 6-34. Find the reactions at A and B.

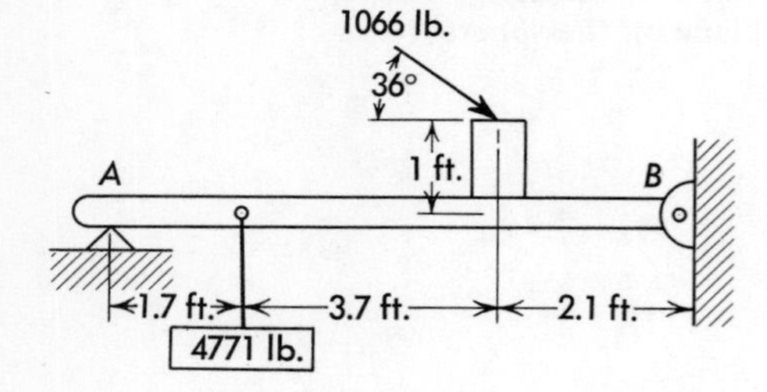

Figure 6-34

6-67. (*a*) Find the tension in AC in Figure 6-35. (*b*) Find the amount and direction of the reaction at B. BC = 10 ft, BD = 25 ft.

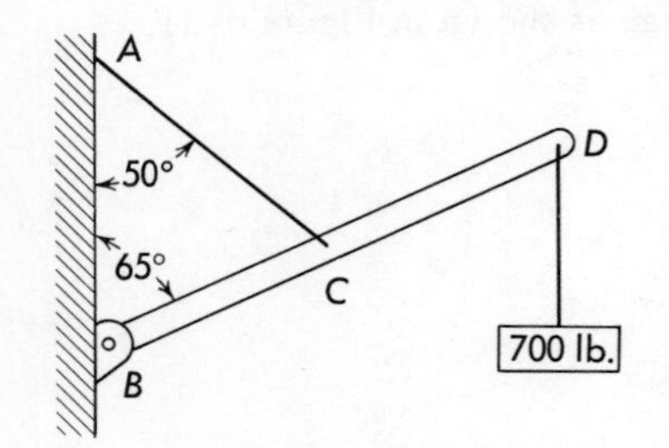

Figure 6-35

6-68. Cylinder No. 1 in Figure 6-36 has a 10-cm diameter and weighs 84 N. Cylinder No. 2 has a 6-cm diameter and weighs 27 N. Find the reactions at *A, B,* and *C.* All surfaces are smooth.

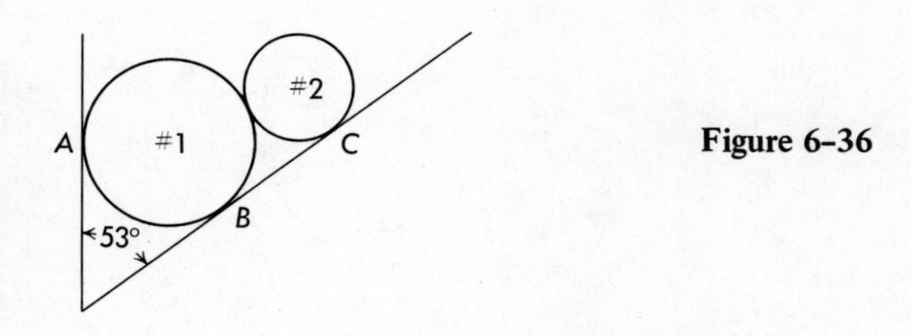

Figure 6-36

6-69. (*a*) Find the force in member *AB* in Figure 6-37 and the reaction at point *E*. (*b*) Find the force in member *CG* and the horizontal and vertical components of the reaction at pin *D*.

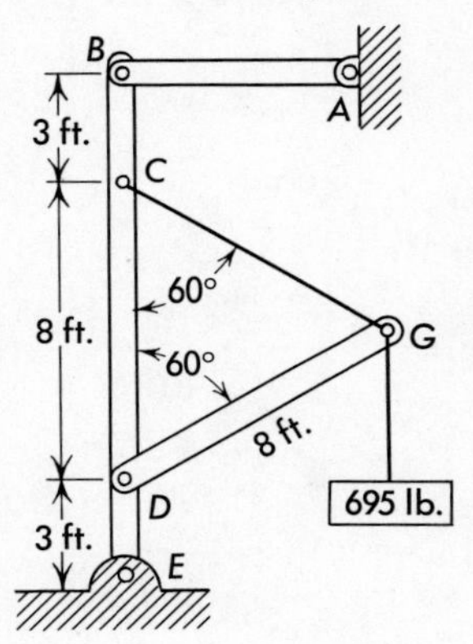

Figure 6-37

Appendix D

logarithms—the mathematical basis for the slide rule

LAWS OF LOGARITHMS

Since a logarithm is an exponent, all the laws of exponents should be reviewed. Let us examine a few of these laws.

Exponential Law I $(a)^m(a)^n = a^{m+n}$

We can put the equation above in statement form, since we know that logarithms are exponents and therefore follow the laws of exponents.

Law I The logarithm of a product equals the sum of the logarithms of the factors.

Example

$$(5)(7) = ?$$
$$\log_{10}5 + \log_{10}7 = \log_{10} \text{ ans.}$$
$$0.6990 + 0.8451 = \log \text{ ans.}$$
$$1.5441 = \log \text{ ans.}$$
$$\text{Answer} = \mathbf{(3.50)(10)^1}$$

This is true because

$$5 = (10)^{0.6990}$$
$$7 = (10)^{0.8451}$$
$$\text{product} = (10)^{0.6990}(10)^{0.8451}$$
$$= (10)^{0.6990+0.8451}$$
$$= (10)^{1.5441}$$
$$= \mathbf{(3.50)(10)^1}$$

Exponential Law II $\frac{a^m}{a^n} = a^{m-n}$

Putting the equation above in statement form, we obtain the following law.

Law II The logarithm of a quotient equals the logarithm of the dividend minus the logarithm of the divisor.

Example

$$\begin{aligned} 5/4 &= ? \\ \log 5 - \log 4 &= \log \text{ans.} \\ 0.6990 - 0.6021 &= \log \text{ans.} \\ 0.0969 &= \log \text{ans.} \\ \text{Answer} &= \mathbf{1.25} \end{aligned}$$

Law III The logarithm of the x power of a number equals x times the logarithm of the number.

Example

$$\begin{aligned} (5)^3 &= ? \\ 3(\log 5) &= \log \text{ans.} \\ 3(0.6990) &= \log \text{ans.} \\ 2.0970 &= \log \text{ans.} \\ \text{Answer} &= \mathbf{(1.25)(10)^2} \end{aligned}$$

Law IV The logarithm of the x root of a number equals the logarithm of the number divided by x.

Example

$$\begin{aligned} \sqrt[3]{3375} &= ? \\ \frac{\log 3375}{3} &= \log \text{ans.} \\ \frac{3.5282}{3} &= \log \text{ans.} \\ 1.1761 &= \log \text{ans.} \\ \text{Answer} &= \mathbf{(1.50)(10)^1} \end{aligned}$$

Note Law IV is actually a special case of Law III.

In some instances a combination of Law III and Law IV may be used.

Example

$$\begin{aligned} (0.916)^{3/4.15} &= ? \\ \frac{(\log 0.916)(3)}{4.15} &= \log \text{ans.} \\ \frac{(9.9619 - 10)(3)}{4.15} &= \log \text{ans.} \end{aligned}$$

Perform multiplication first:

$$\frac{29.8857 - 30}{4.15} = \log \text{ans.}$$

To be divided by 4.15, the negative number must be divisible a whole number of times. Therefore, the characteristic (which is -1) is written as $414.0000 - 415$. There are several values which could be chosen, such as 4149.0000 − 4150, which would satisfy the condition that the characteristic be −1. Rewriting and dividing,

$$\frac{414.8857 - 415}{4.15} = \log \text{ ans.}$$

$$99.9725 - 100 = \log \text{ ans.}$$

$$\text{Answer} = \mathbf{(9.39)(10)^{-1}}$$

The cologarithm Many times it is helpful to use the cologarithm of a number rather than the logarithm. The cologarithm of a number is the logarithm of the reciprocal of the number. The cologarithm is also the difference between the logarithm and the logarithm of unity.

Example

$$\begin{aligned} \text{colog } 5 &= \log \frac{1}{5} \\ &= \log 1 - \log 5 \\ &= 0.0000 - 0.6990 \\ &= \mathbf{-0.6990} \end{aligned}$$

Since log 5 equals 0.6990, we see that the colog $x = -\log x$. Therefore:

1. The logarithm of the quotient of two numbers equals the logarithm of the dividend plus the cologarithm of the divisor.
2. The logarithm of the product of two numbers equals the logarithm of one number minus the cologarithm of the other number.

Natural logarithms When certain derivations of engineering formulas are made, a term may appear that contains a natural logarithm. For example, the magnetic field intensity near a current-carrying conductor varies with distance from the conductor according to a logarithmic pattern. In advanced texts it may be shown that a natural logarithm function, when plotted, gives an exponential curve whose slope at any point is equal to the ordinate at that point.

In solving problems involving natural logarithms, tables of natural logarithms can be used if they are available, or the natural logarithm, frequently abbreviated as "ln," may be converted to a logarithm to the base 10. To perform this latter operation, an algebraic transformation called *change of logarithmic base* is used. This transformation can be performed as follows:

$$\text{Natural logarithm} = (\text{common log})(\log_\epsilon 10)$$

Since $\log_\epsilon 10 = 2.3026$, we may write:

$$\text{Natural logarithm} = (\text{common log})(2.3026)$$

If natural logarithms are computed, it must be remembered that the mantissa is not independent of the location of the decimal point. Therefore, the same sequence of significant figures does not have the same mantissa, as is the case with common logarithms.

Example Find the natural logarithm of 245.

$$\begin{aligned} \log_{10} 245 &= 2.3892 \\ \ln 245 &= (2.3892)(2.3026) \\ &= \mathbf{5.5014} \end{aligned}$$

Example Find the natural logarithm of 2.45.

$$\begin{aligned} \log_{10} 245 &= 0.3892 \\ \ln 2.45 &= (0.3892)(2.3026) \\ &= \mathbf{0.8961} \end{aligned}$$

The natural logarithm of a number less than 1 is a negative number.

Example Find the natural logarithm of 0.245.

$$\log_{10} 0.245 = 9.3892 - 10$$

Since the logarithm has a negative characteristic, we can solve by first finding the colog and then multiplying by $\log_{\epsilon} 10$.

$$\begin{aligned} \text{colog}_{10}\, 0.245 &= -0.6108 \\ \ln 0.245 &= (-0.6108)(2.3026) \\ &= \mathbf{-1.4064} \end{aligned}$$

Appendix II

trigonometry

RIGHT TRIANGLES

It can be shown by measurements and by formal derivations that for any given size of an angle at A or C, the ratio of the lengths of the sides to each other in a right triangle is a constant regardless of the numerical value of the lengths. In Figure AII-1, the sides of a right triangle are named in reference to the angle under consideration. In the cases, the angle is designated as θ (theta).

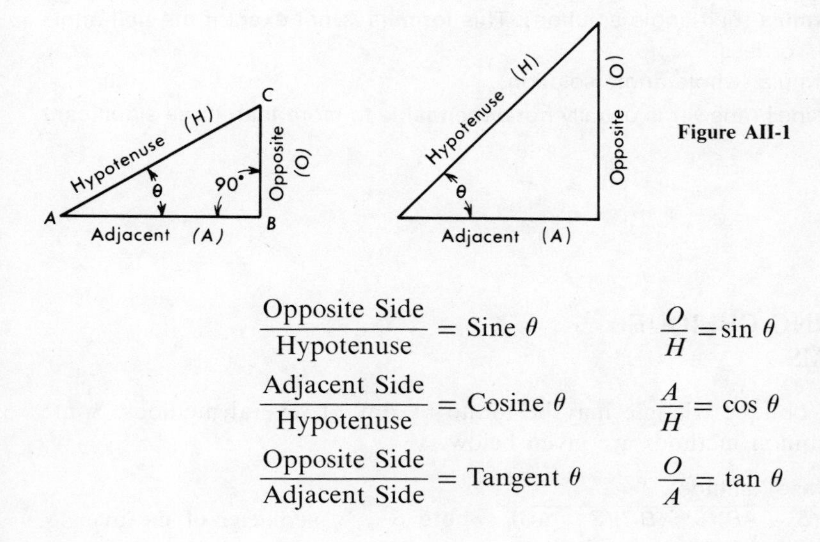

Figure AII-1

$$\frac{\text{Opposite Side}}{\text{Hypotenuse}} = \text{Sine } \theta \qquad \frac{O}{H} = \sin \theta$$

$$\frac{\text{Adjacent Side}}{\text{Hypotenuse}} = \text{Cosine } \theta \qquad \frac{A}{H} = \cos \theta$$

$$\frac{\text{Opposite Side}}{\text{Adjacent Side}} = \text{Tangent } \theta \qquad \frac{O}{A} = \tan \theta$$

$$\frac{\text{Adjacent Side}}{\text{Opposite Side}} = \text{Cotangent } \theta \qquad \frac{A}{O} = \cot \theta$$

$$\frac{\text{Hypotenuse}}{\text{Adjacent Side}} = \text{Secant } \theta \qquad \frac{H}{A} = \sec \theta$$

$$\frac{\text{Hypotenuse}}{\text{Opposite Side}} = \text{Cosecant } \theta \qquad \frac{H}{O} = \csc \theta$$

METHODS FOR FINDING AREAS OF OBLIQUE TRIANGLES

In order to solve an oblique triangle problem, at least three of the six parts of the triangle must be known, and at least one of the known parts must be a side. In the suggested methods listed below, only the most effective methods are given.

1. Given: two sides and an angle opposite one of them:
   a. Law of sines.
   b. Right triangles.

2. Given: two angles and one side:
   a. Law of sines.
   b. Right triangles.

3. Given: two sides and the included angle:
   a. Law of cosines (answer is usually not dependable to more than three significant figures).
   b. Right triangles.

4. Given: three sides only:
   a. Tangent formula (half-angle solution).
   b. Sine formula (half-angle solution). This formula is not exact if the half-angle is near 90°.
   c. Cosine formula (half-angle solution). This formula is not exact if the half-angle is about 6° or less.
   d. Cosine formula (whole angle solution).
   e. Law of cosines (answer is usually not dependable to more than three significant figures).

METHODS OF SOLVING OBLIQUE TRIANGLE PROBLEMS

The area of an oblique triangle may be found by any of several methods. Some of the more common methods are given below:

1. Area = ($\frac{1}{2}$)(base)(altitude).
2. Area = $\sqrt{(S)(S - AB)(S - BC)(S - AC)}$, where $S = \frac{1}{2}$ perimeter of the triangle.
3. Area = $\frac{1}{2}$ (product of two sides) (sine of the included angle).

SINE LAW

In any triangle the ratio of the length of a side to the sine of the angle opposite that side is the same as the ratio of any other side to the sine of the angle opposite it. In symbol form (see Figure AII-2):

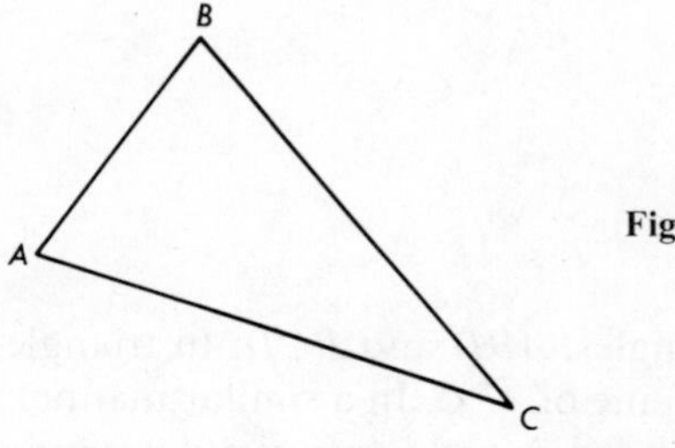

Figure AII-2

$$\frac{AB}{\sin \sphericalangle C} = \frac{BC}{\sin \sphericalangle A} = \frac{AC}{\sin \sphericalangle B}$$

This expression is called the *sine law*. The student is cautioned not to confuse the meanings of sine functions and sine law.

In the event one of the angles of a triangle is larger than 90°, a simple way to obtain the value of the sine of the angle is to subtract the angle from 180° and obtain the sine of this angle to use in the sine law expression.

The sine law can also be used if two sides and an angle of a triangle are known, provided the angle is not the one included between the sides. However, as explained in trigonometry texts, the product of the sine of the angle and the side adjacent must be equal to or less than the side opposite the angle; otherwise no solution is possible.

As an alternate method, the general triangle can be made into right triangles by adding construction lines. This method of using right triangle solutions is as exact as the sine law but usually will take more time than the sine law method.

COSINE LAW

In an oblique triangle, the square of any side is equal to the sum of the squares of the other two sides minus twice the product of the other two sides times the cosine of the included angle. In symbol form:

$$(AB)^2 = (AC)^2 + (BC)^2 - (2)(AC)(BC)(\cos \sphericalangle C)$$

This expression is called the *cosine law* and is useful in many problems, although it may not give an answer to the desired precision since we are adding and subtracting terms that have only three significant figures.

After the side AB has been determined, the angles at A and B can be found by using the law of sines.

In the event that the angle used in the cosine law formula is larger than 90°, subtract the angle from 180°, and determine the cosine of this angle. Remember, however, that the cosine of an angle between 90° and 180° is negative. If the angle used in the formula is larger than 90°, the last term will add to the squared terms.

The problem above can also be solved by using construction lines and making right triangles from the figure (Figure AII-3). To do this, we construct the line BD

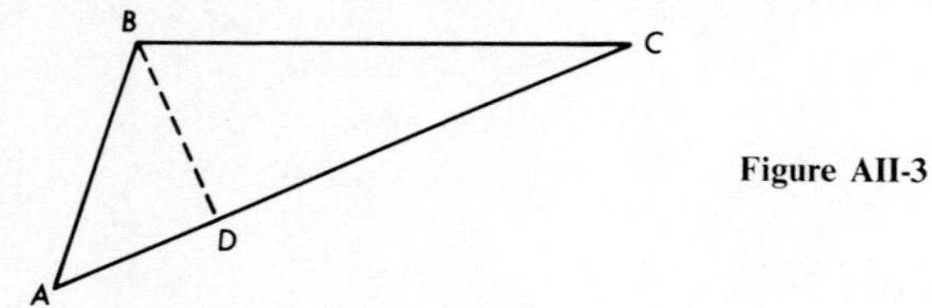

Figure AII-3

perpendicular to AC. This will form two right triangles, ABD and BCD. In triangle BCD, side BD may be found by using BC and the sine of $\measuredangle C$. In a similar manner, by using the cosine of $< C$, side DC may be found. From this we can determine side AD in triangle ABD.

Using the tangent function, the angle at A can be found, and AB can be determined by the use of the sine or cosine function or the Pythagorean theorem $(AB)^2 = (BD)^2 + (AD)^2$. The right triangle method, while it may take longer to solve, will in general give a more accurate answer.

THREE SIDES LAWS

There are a number of formulas derived in trigonometry that will give the angles of an oblique triangle when only three sides are known. The formulas differ considerably in ease of application and precision, especially if logarithms are used. Of all the formulas available, in general the half-angle (tangent) formula is better than others. The formula (half-angle solution) is as follows:

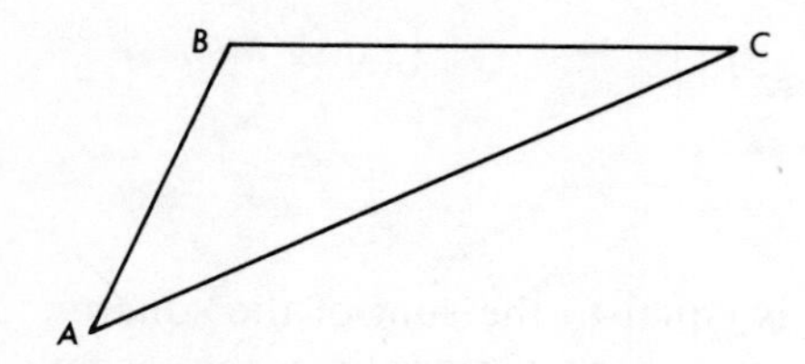

Figure AII-4

$$\tan \tfrac{1}{2} A = \frac{r}{S - BC}$$

where
$$r = \sqrt{\frac{(S - AB)(S - AC)(S - BC)}{S}}$$

and
$$S = \tfrac{1}{2} \text{ perimeter of triangle}$$

Other formulas that may be used are the following:

$$\text{Sine formula (half-angle solution) } \sin \tfrac{1}{2} A = \sqrt{\frac{(S - AC)(S - AB)}{(AC)(AB)}}$$

$$\text{Cosine formula (half-angle solution) } \cos \tfrac{1}{2} A = \sqrt{\frac{(S)(S - BC)}{(AC)(AB)}}$$

$$\text{Cosine formula (whole angle solution) } \cos A = \frac{(2S)(S - BC)}{(AB)(AC)} - 1$$

In the last formula, the quantity $(2S)(S - BC)/(AB)(AC)$ will usually be between 1 and 2 and can be read to four figures on the slide rule. Subtracting the 1 in the equation will leave the cosine of the angle correct to three figures. The formula has the advantage that it requires fewer operations. Also it is convenient to use if the slide rule is employed in solving problems.

After finding one angle, the remaining angles can be found by successive applications of the law, being careful to use the proper side of the triangle in the formula. The sine law can also be used after one angle is found. In order to have a check on the solution, it is better to solve for all three angles rather than solve for two angles and then subtract their sum from 180°. If each angle is computed separately, their sum should be within the allowable error range of 180°.

As an incidental item in the tangent formula, the constant r is equal to the length of the radius of a circle that can be inscribed in the triangle.

Appendix III

geometric figures

Rectangle

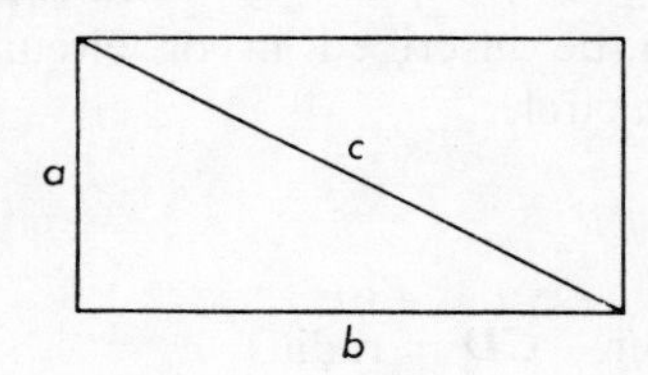

Area = (base) (altitude) = ab

Diagonal = $\sqrt{(\text{altitude})^2 + (\text{base})^2}$

$$C = \sqrt{a^2 + b^2}$$

Right triangle

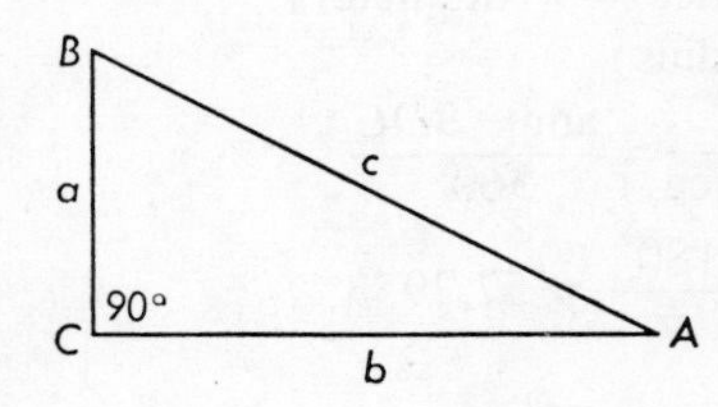

Angle A + angle B = angle C = 90°

Area = ½ (base) (altitude)

Hypotenuse = $\sqrt{(\text{altitude})^2 + (\text{base})^2}$

$$C = \sqrt{a^2 + b^2}$$

Any triangle

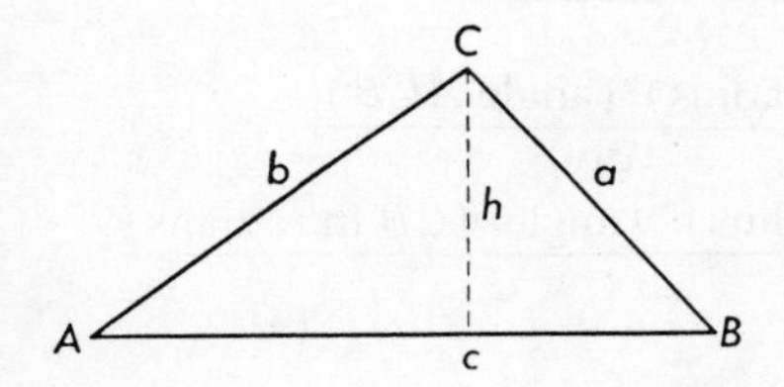

Angles $A + B + C = 180°$

(Altitude h is perpendicular to base c)

Area = ½ (base) (altitude)

Parallelogram

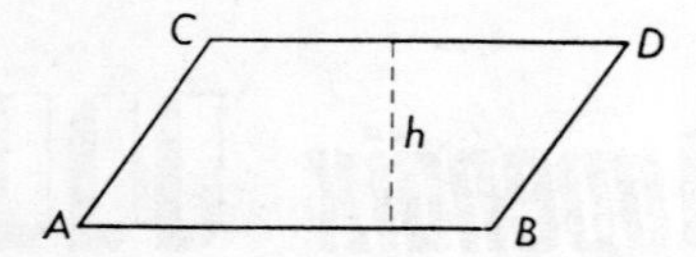

Area = (base) (altitude)
Altitude h is perpendicular to base AB
Angles $A + B + C + D = 360°$

Trapezoid

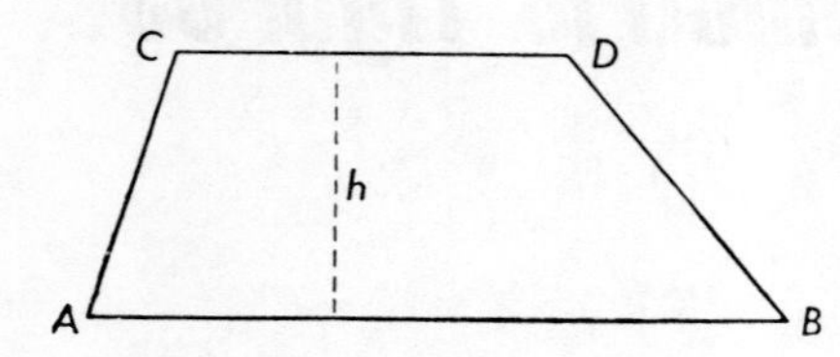

Area = ½ (altitude) (sum of bases)
(Altitude h is perpendicular to sides AB and CD. Side AB is parallel to side CD.)

Regular polygon

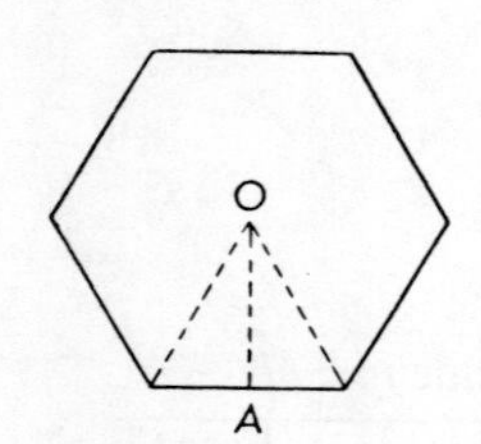

$$\text{Area} = \tfrac{1}{2}\begin{bmatrix}\text{length of}\\ \text{one side}\end{bmatrix}\begin{bmatrix}\text{number}\\ \text{of sides}\end{bmatrix}\begin{bmatrix}\text{distance}\\ OA \text{ to}\\ \text{center}\end{bmatrix}$$

A regular polygon has equal angles and equal sides and can be inscribed in or circumscribed about a circle.

Circle

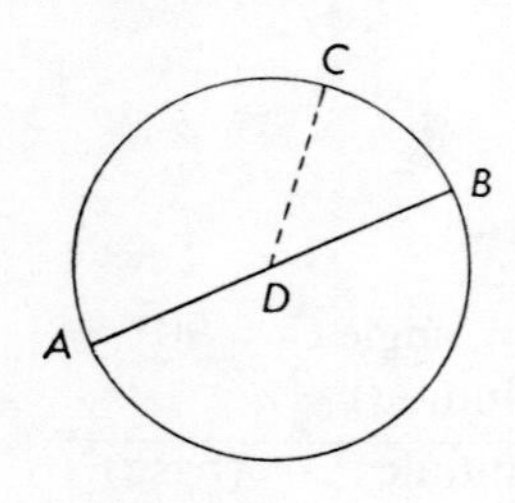

AB = diameter, CD = radius

$$\text{Area} = \pi(\text{radius})^2 = \frac{\pi(\text{diameter})^2}{4}$$

Circumference = π(diameter)

$C = 2\pi(\text{radius})$

$$\frac{\text{arc } BC}{\text{circumference}} = \frac{\text{angle } BDC}{360°}$$

$$1 \text{ radian} = \frac{180°}{\pi} = 57.2958°$$

Sector of a circle

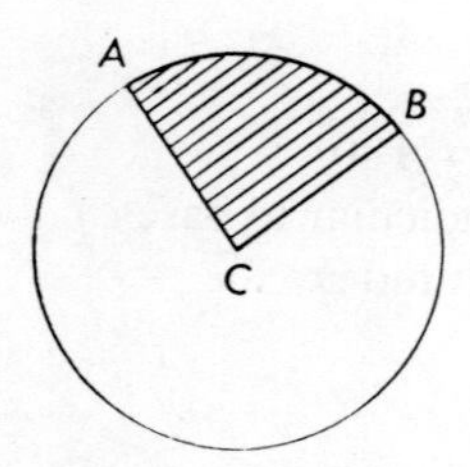

$$\text{Area} = \frac{(\text{arc } AB)(\text{radius})}{2}$$

$$= \pi\frac{(\text{radius})^2(\text{angle } ACB)}{360°}$$

$$= \frac{(\text{radius})^2\ (\text{angle } ACB \text{ in radians})}{2}$$

Segment of a circle

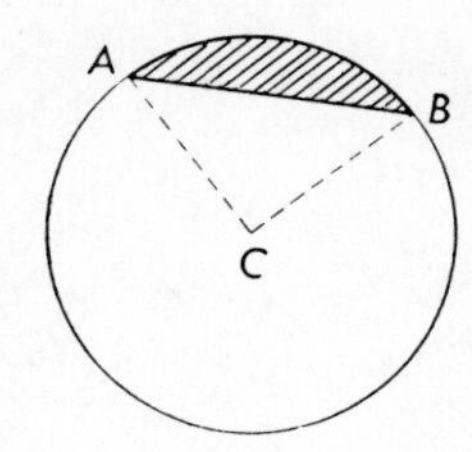

$$\text{Area} = \frac{(\text{radius})^2}{2}\left[\frac{\pi(\measuredangle ACB^\circ)}{180} - \sin ACB^\circ\right]$$

$$\text{Area} = \frac{(\text{radius})^2}{2}\left[\measuredangle ACB \text{ in radians} - \sin ACB^\circ\right]$$

Area = area of sector ACB − area of triangle ABC

Ellipse

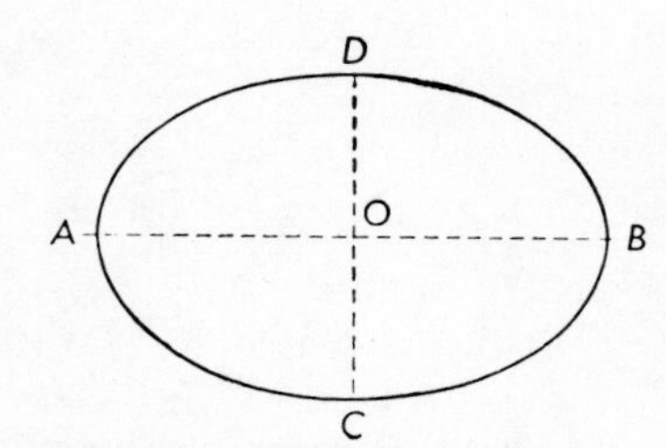

Area = π(long radius OA)(short radius OC)

Area = $\frac{\pi}{4}$ (long diameter AB) (short diameter CD)

Rectangular prism

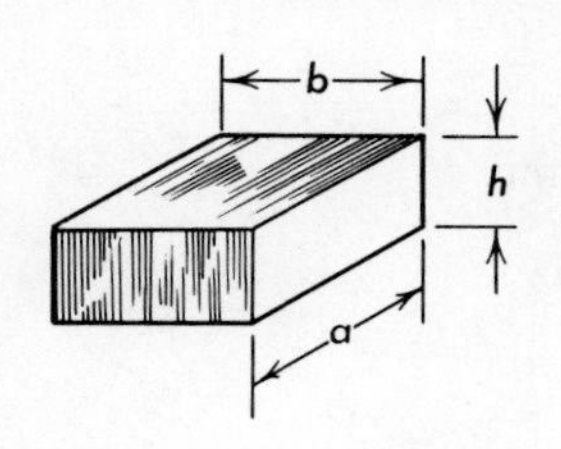

Volume = length × width × height
Volume = area of base × altitude

Any prism

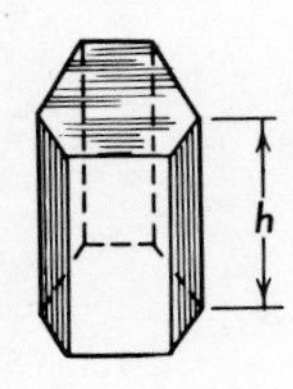

(Axis either perpendicular or inclined to base)
Volume = (area of base) (perpendicular height)
Volume = (lateral length) (area of perpendicular cross section)

Volume and center of gravity equations[1]

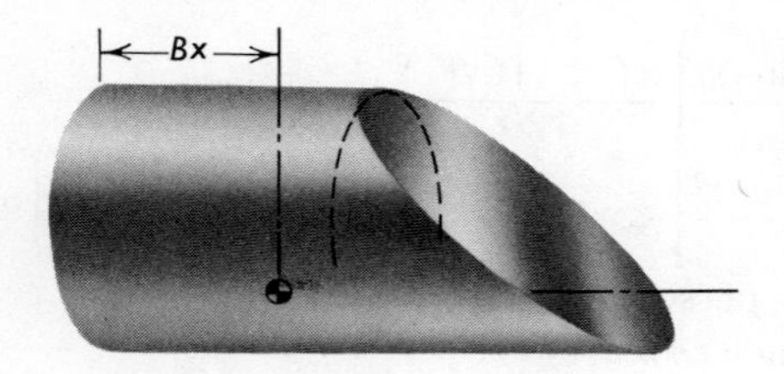

Volume equations are included for all cases. Where the equation for the CG (center of gravity) is not given, you can easily obtain it by looking up the volume and CG equations for portions of the shape and then combining values. For example, for the shape above, use the equations for a cylinder, Figure 1, and a truncated cylinder, Figure 10 (subscripts C and T, respectively, in the equations below). Hence taking moments,

$$B_x = \frac{V_C B_C + V_T(B_T + L_C)}{V_C + V_T}$$

or $$B_x = \frac{\left(\frac{\pi}{4} D^2 L_C\right)\left(\frac{L_C}{2}\right) + \frac{\pi}{8} D^2 L_T\left(\frac{5}{16} L_T + L_C\right)}{\frac{\pi}{4} D^2 L_C + \frac{\pi}{8} D^2 L_T}$$

$$B_x = \frac{L^2_C + L_T\left(\frac{5}{16} L_T + L_C\right)}{2L_C + L_T}$$

[1] Courtesy of Knoll Atomic Power Laboratory, Schenectady, New York, operated by the General Electric Company for the U.S. Atomic Energy Commission. Reprinted from *Product Engineering*—Copyright owned by McGraw-Hill, New York.

In the equations to follow, angle θ can be either in degrees or in radians.

Thus θ (rad) $= \pi\theta/180$ (deg) $= 0.01745\ \theta$ (deg).

For example, if $\theta = 30$ deg in Case 3, then $\sin\theta = 0.5$ and

$$B = \frac{2R\ (0.5)}{3\ (30)\ (0.01745)} = 0.637R$$

Symbols used are:

B = distance from CG to reference plane,

V = volume,

D and d = diameter,

R and r = radius,

H = height,

L = length.

1. Cylinder

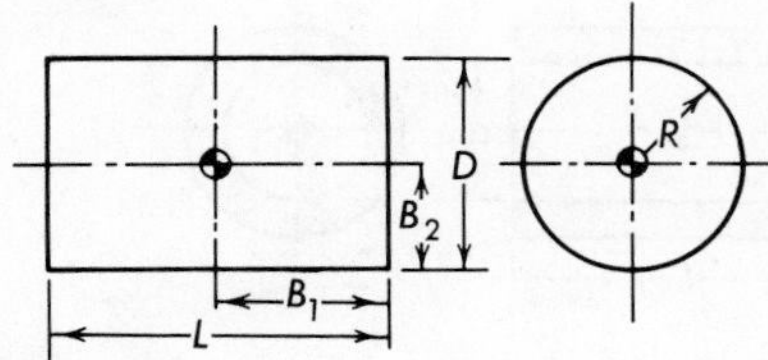

$$V = \frac{\pi}{4}D^2L = 0.7854D^2L \qquad B_1 = L/2$$
$$B_2 = R$$

Area of cylindrical surface
= (Perimeter of base) (perpendicular height)

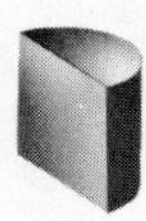

2. Half cylinder

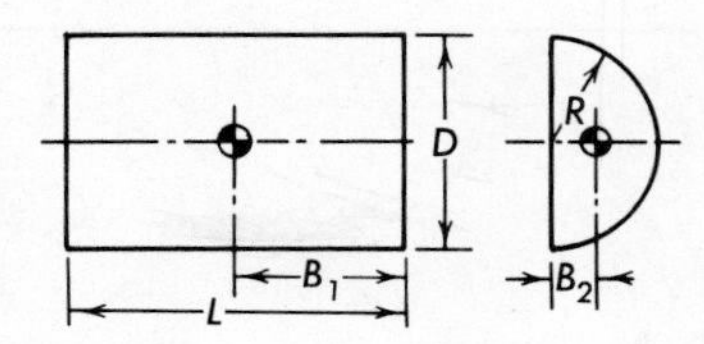

$$V = \frac{\pi}{8}D^2L = 0.3927D^2L$$

$$B_1 = L/2 \qquad B_2 = \frac{4R}{3\pi} = 0.4244R$$

3. Sector of cylinder

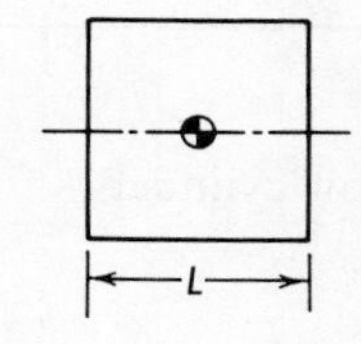

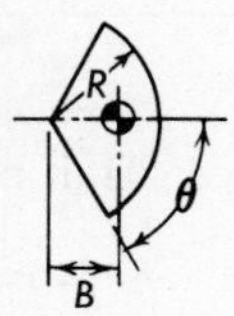

$$V = \theta R^2L \qquad B = \frac{2R\sin\theta}{3\theta}$$

4. Segment of cylinder

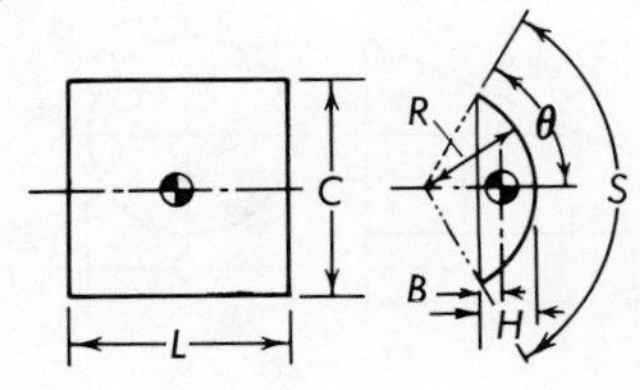

$$V = LR^2\left(\theta - \frac{1}{2}\sin 2\theta\right)$$
$$V = 0.5L\,[RS - C\,(R - H)]$$
$$B = \frac{4R\sin^3\theta}{6\theta - 3\sin 2\theta}$$
$$S = 2R\theta$$
$$H = R\,(1 - \cos\theta)$$
$$C = 2R\sin\theta$$

5. Quadrant of cylinder

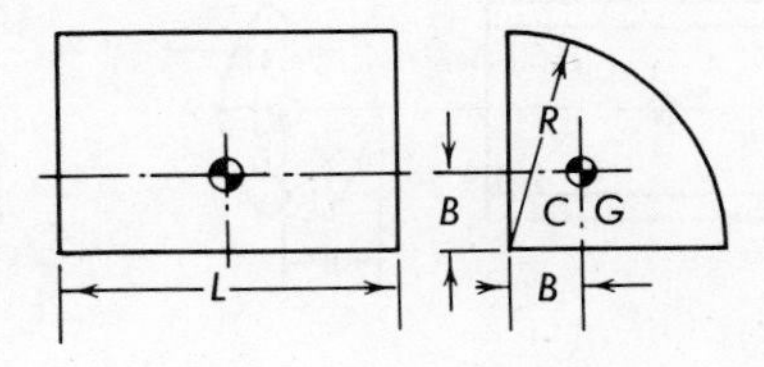

$$V = \frac{\pi}{4}R^2L = 0.7854R^2L$$

$$B = \frac{4R}{3\pi} = 0.4244R$$

6. Fillet or spandrel

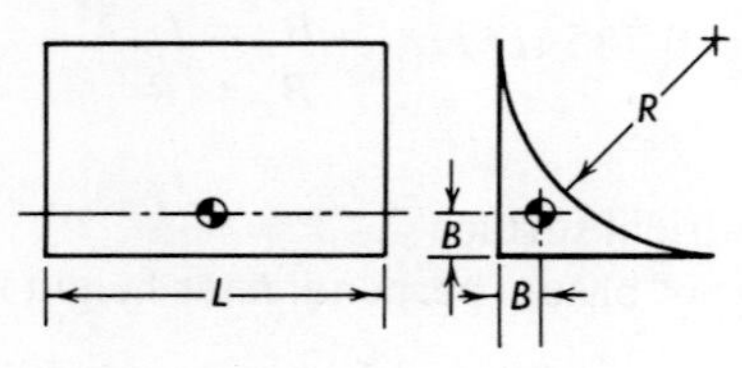

$$V = \left(1 - \frac{\pi}{4}\right) R^2L = 0.2146R^2L$$

$$B = \frac{10 - 3\pi}{12 - 3\pi} R = 0.2234R$$

7. Hollow cylinder

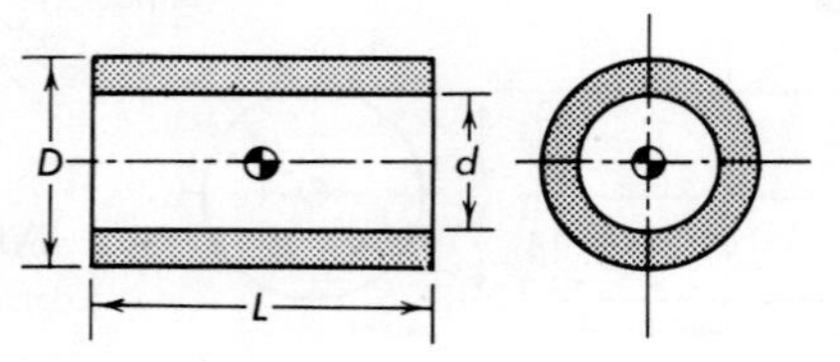

$$V = \frac{\pi L}{4}(D^2 - d^2)$$

CG at center of part

8. Half hollow cylinder

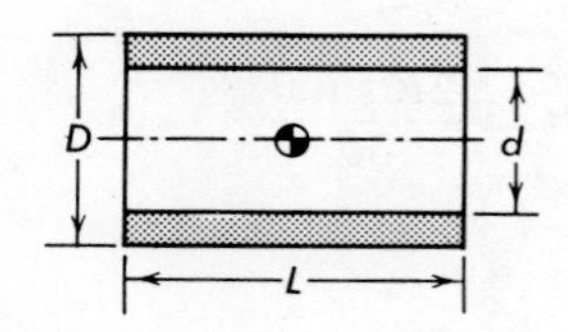

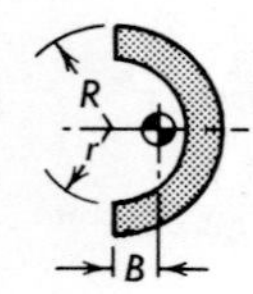

$$V = \frac{\pi L}{8}(D^2 - d^2)$$

$$B = \frac{4}{3\pi}\left[\frac{R^3 - r^3}{R^2 - r^2}\right]$$

9. Sector of hollow cylinder

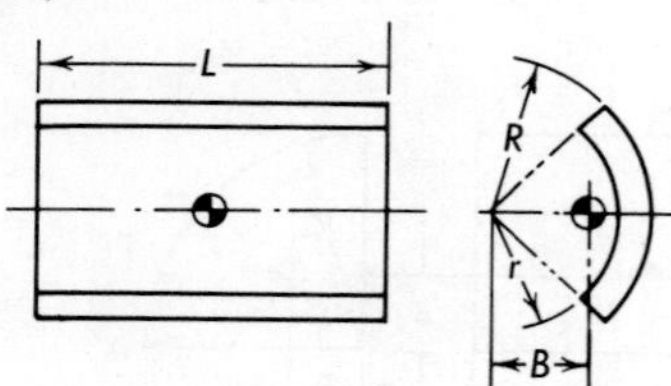

$$V = 0.01745\ (R^2 - r^2)\ \theta L$$

$$B = \frac{38.1972\ (R^3 - r^3)\ \sin\theta}{(R^2 - r^2)\ \theta}$$

10. Truncated cylinder (with full circle base)

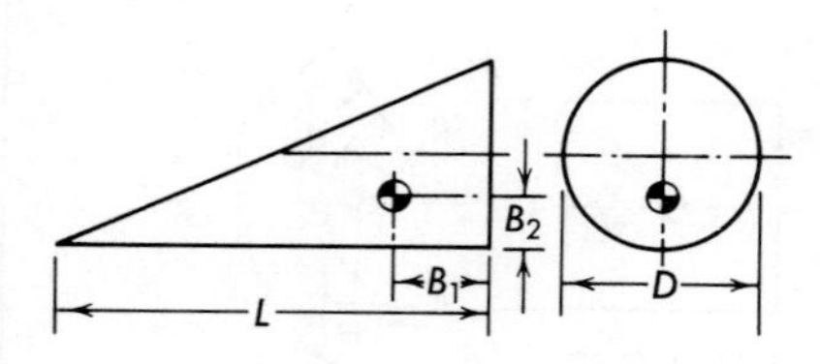

$$V = \frac{\pi}{8}D^2L = 0.3927D^2L$$

$$B_1 = 0.3125L$$

$$B_2 = 0.375D$$

11. Truncated cylinder (with partial circle base)

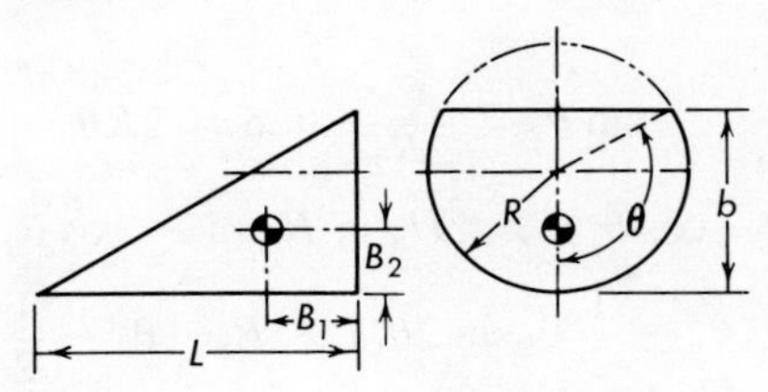

$$b = R\,(1 - \cos\theta)$$

$$V = \frac{R^3L}{b}\left[\sin\theta - \frac{\sin^3\theta}{3} - \theta\cos\theta\right]$$

$$B_1 = \frac{L\left[\frac{\theta\cos^2\theta}{2} - \frac{5\sin\theta\cos\theta}{8} + \frac{\sin^3\theta\cos\theta}{12} + \frac{\theta}{8}\right]}{\left[1 - \cos\theta\right]\left[\sin\theta - \frac{\sin^3\theta}{3} - \theta\cos\theta\right]}$$

$$B_2 = \frac{2R\left[-\frac{\theta\cos\theta}{2} + \frac{\sin\theta}{2} - \frac{\theta}{8} + \frac{\sin\theta\cos\theta}{8} - N\right]}{\left[\sin\theta - \frac{\sin^3\theta}{3} - \theta\cos\theta\right]}$$

$$\text{where } N = \frac{\sin^3\theta}{6} - \frac{\sin^3\theta\cos\theta}{12}$$

12. Oblique cylinder
(or circular hole at oblique angle)

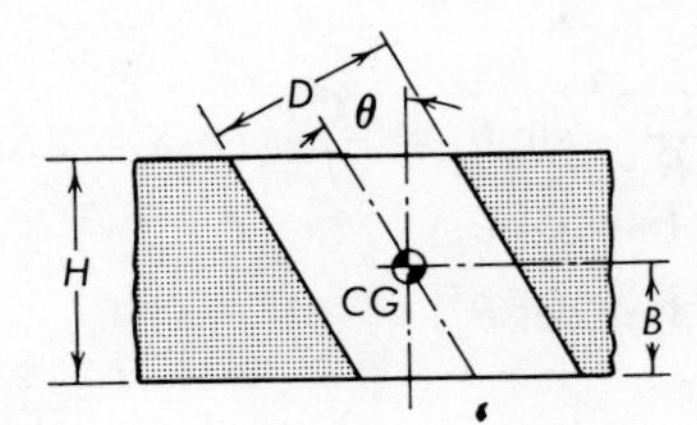

$$V = \frac{\pi}{4}D^2\frac{H}{\cos\theta} = 0.7854D^2H\sec\theta$$

$$B = H/2 \qquad r = \frac{d}{2}$$

13. Bend in cylinder

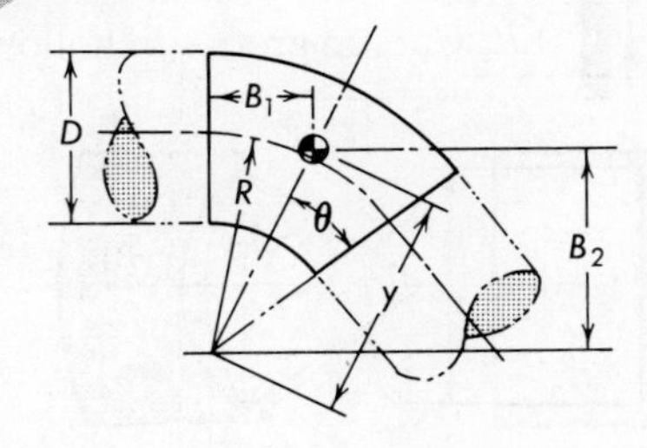

$$V = \frac{\pi^2}{360}D^2R\theta = 0.0274D^2R\theta$$

$$y = R\left[1 + \frac{r^2}{4R^2}\right] \qquad B_1 = y\tan\theta \quad B_2 = y\cot\theta$$

14. Curved groove in cylinder

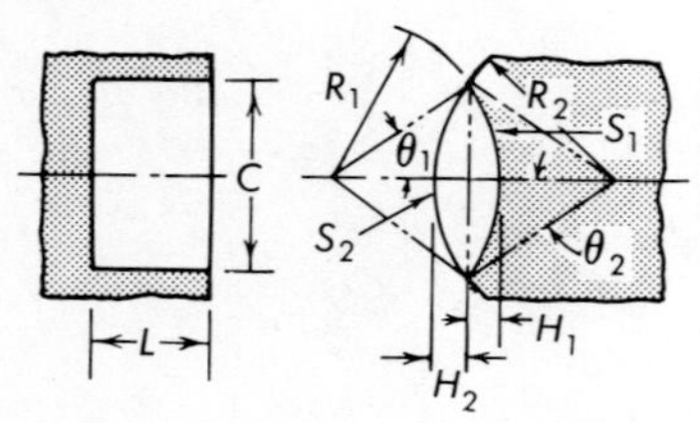

$$\sin\theta_1 = \frac{C}{2R_1} \qquad \sin\theta_2 = \frac{C}{2R_2} \qquad S = 2R\theta$$

$$H_1 = R_1\,(1 - \cos\theta_1) \qquad H_2 = R_2\,(1 - \cos\theta_2)$$

$$V = L\left[R_1^2\,(\theta_1 - \frac{1}{2}\,\theta_1 \sin 2\theta_1) + R_2^2\,(\theta_2 - \frac{1}{2}\,\theta_2 \sin 2\theta\right.$$

Compute CG of each part separately

15. Slot in cylinder

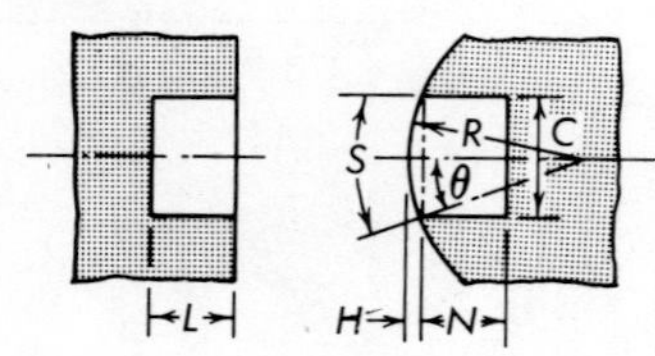

$$H = R\,(1 - \cos\theta) \qquad \sin\theta = \frac{C}{2R}$$

$$S = 2R\theta$$

$$V = L\left[CN + R^2\,(\theta - \frac{1}{2}\sin 2\theta)\right]$$

16. Slot in hollow cylinder

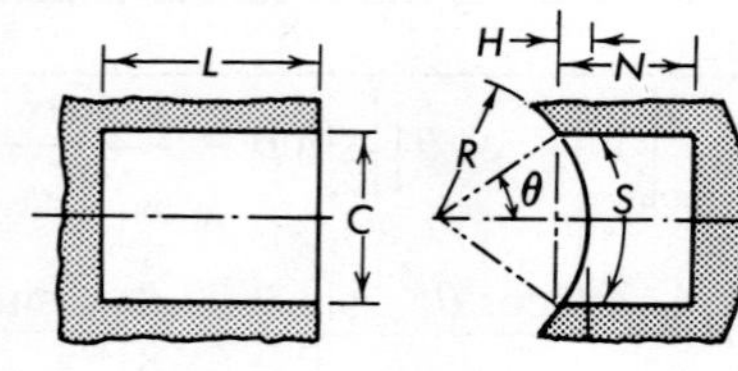

$$S = 2R\theta \qquad \sin\theta = \frac{C}{2R}$$

$$H = R\,(1 - \cos\theta)$$

$$V = L\left[CN - R^2\,(\theta - \frac{1}{2}\sin 2\theta)\right]$$

$$V = L\left\{CN - 0.5\,[RS - C\,(R - H)]\right\}$$

17. Curved groove in hollow cylinder

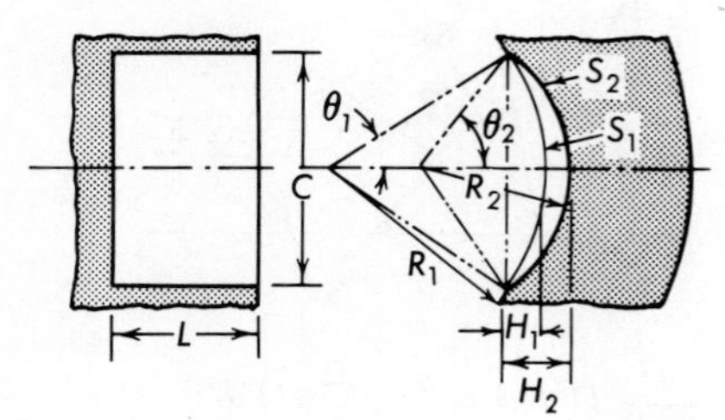

$$\sin\theta_1 = \frac{C}{2R_1} \qquad \sin\theta_2 = \frac{C}{2R_2} \qquad S = 2R\theta$$

$$H_1 = R_1\,(1 - \cos\theta_1)$$

$$H_2 = R_2\,(1 - \cos\theta_2)$$

$$V = L\left(\left[R_2^2\,(\theta_2 - \frac{1}{2}\sin 2\theta_2)\right] - \left[R_1^2\,(\theta_1 - \frac{1}{2}\sin 2\theta_1)\right]\right)$$

$$V = \frac{L}{2}\left(\left[R_2S_2 - C\,(R_2 - H_2)\right] - \left[R_1S_1 - C\,(R_1 - H_1)\right]\right)$$

18. Slot through hollow cylinder

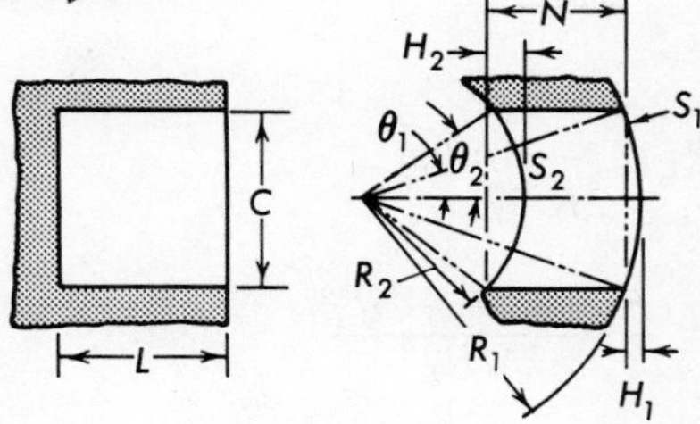

$$\sin \theta_1 = \frac{C}{R_1} \qquad \sin \theta_2 = \frac{C}{R_2}$$

$$S = 2R\theta$$

$$H_1 = R_1\,(1 - \cos \theta_1)$$

$$H_2 = R_2\,(1 - \cos \theta_2)$$

$$V = L\left(CN + \left[R_1^2\,(\theta_1 - \frac{1}{2}\sin 2\theta_1)\right] - \left[R_2\,(\theta_2 - \frac{1}{2}\sin \theta_2)\right]\right)$$

$$V = L\left(CN + 0.5\,[R_1S_1 - C\,(R_1 - H_1)] - 0.5\,[R_2S_2 - C\,(R_2 - H_2)]\right)$$

19. Intersecting cylinder (volume of junction box)

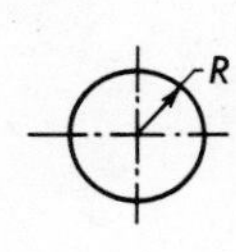

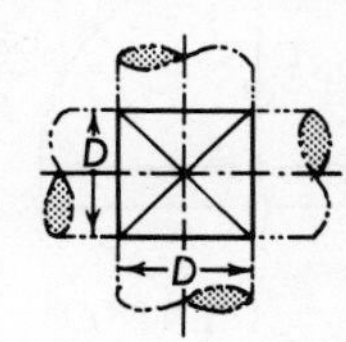

$$V = D^3\left(\frac{\pi}{2} - \frac{2}{3}\right) = 0.9041D^3$$

20. Intersecting hollow cylinders (volume of junction box)

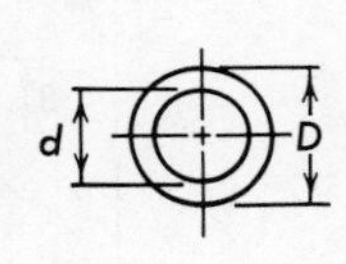

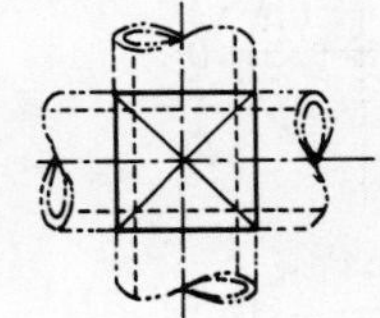

$$V = \left(\frac{\pi}{2} - \frac{2}{3}\right)(D^3 - d^3) - \frac{\pi}{2}d^2\,(D - d)$$

$$V = 0.9041\,(D^3 - d^3) - 1.5708\,d^2\,(D - d)$$

21. Intersecting parallel cylinders ($M < R_1$)

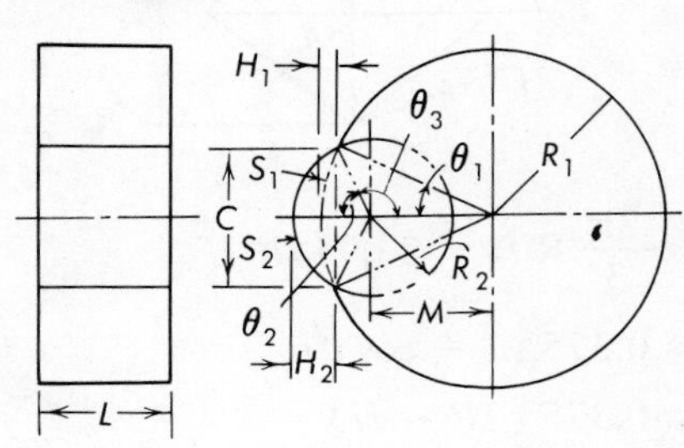

$$\theta_2 = 180^\circ - \theta_3 \qquad \cos \theta_3 = \frac{R_2^2 + M^2 - R_1^2}{2MR_2}$$

$$\cos \theta_1 = \frac{R_1^2 + M^2 - R_2^2}{2MR_1}$$

$$H_1 = R_1\,(1 - \cos \theta_1)$$

$$S_1 = 2R_1\theta_1$$

$$V = L\left(\pi R_1^2 + \left[R_2^2\,(\theta_2 - \frac{1}{2}\sin 2\theta_2)\right] - \left[R_1^2\,(\theta_1 - \frac{1}{2}\sin 2\theta_1)\right]\right)$$

22. Intersecting parallel cylinders ($M > R_1$)

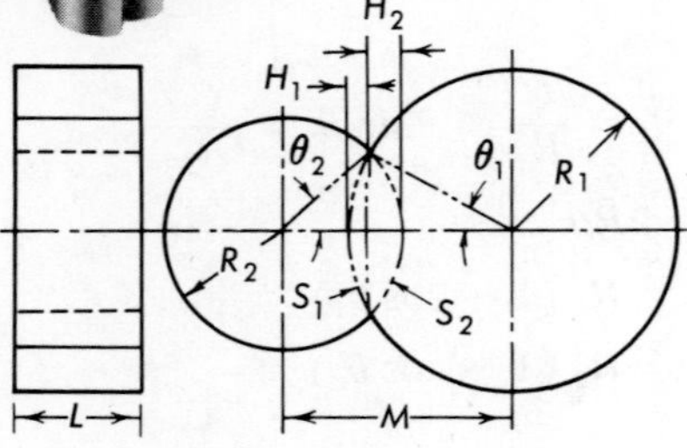

$$H_1 = R_1 (1 - \cos \theta_1)$$
$$S_1 = 2R_1\theta_1$$

$$\cos \theta_1 = \frac{R_1^2 + M^2 - R_2^2}{2MR_1}$$

$$V = L\left(\left[\pi (R_1^2 + R_2^2)\right] - \left[R_1^2 \left(\theta_1 - \frac{1}{2} \sin 2\theta_1\right)\right] - \left[R_2^2 \left(\theta_2 - \frac{1}{2} \sin 2\theta_2\right)\right]\right)$$

23. Sphere

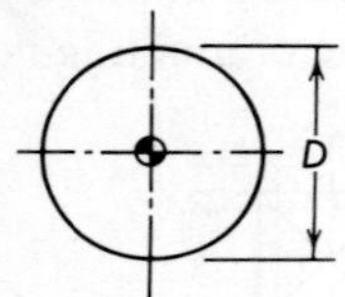

$$V = \frac{\pi D^3}{6} = 0.5236D^3$$

Area of surface $= 4\pi(\text{radius})^2 = \pi D^2$

24. Hemisphere

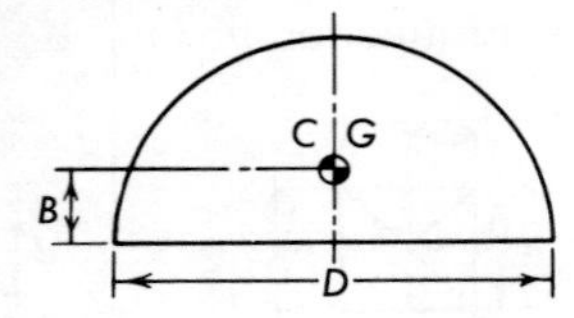

$$V = \frac{\pi D^3}{12} = 0.2618D^3$$

$B = 0.375R$

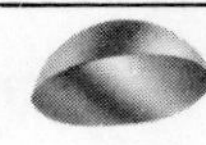

25. Spherical segment

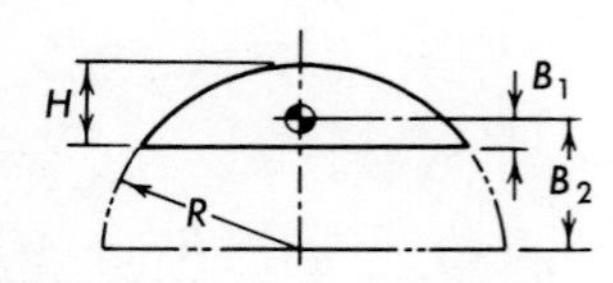

$$V = \pi H^2\left(R - \frac{H}{3}\right)$$

$$B_1 = \frac{H (4R - H)}{4 (3R - H)}$$

$$B_2 = \frac{3 (2R - H)^2}{4 (3R - H)}$$

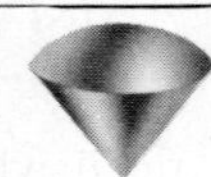

26. Spherical sector

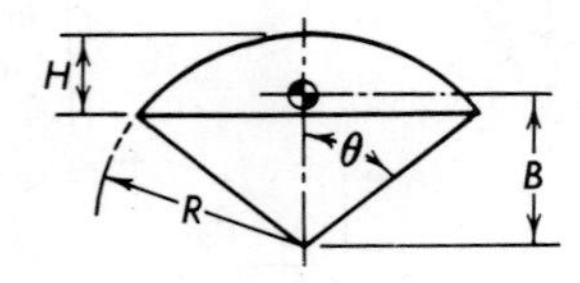

$$V = \frac{2\pi}{3} R^2H = 2.0944R^2H$$

$B = 0.375 (1 + \cos \theta)$

$R = 0.375 (2R - H)$

27. Shell of hollow hemisphere

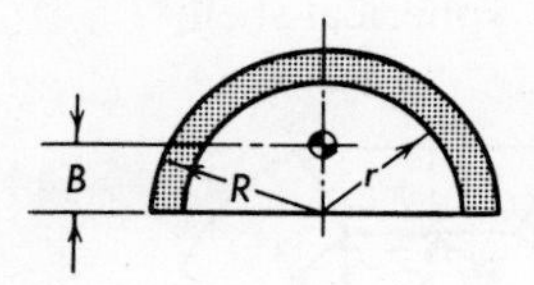

$$V = \frac{2\pi}{3}(R^3 - r^3)$$

$$B = 0.375\left(\frac{R^4 - r^4}{R^3 - r^3}\right)$$

28. Hollow sphere

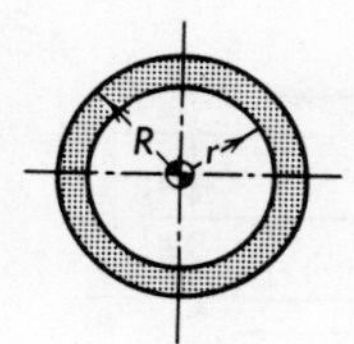

$$V = \frac{4\pi}{3}(R^3 - r^3)$$

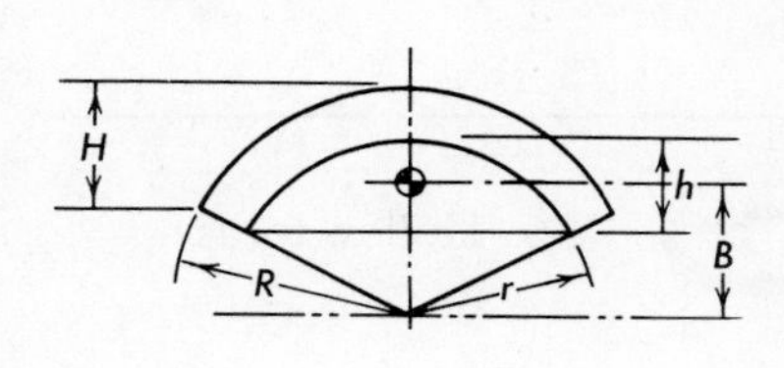

29. Shell of spherical sector

$$V = \frac{2\pi}{3}(R^2H - r^2h)$$

$$B = 0.375\left\{\frac{[R^2H\,(2R - H)] - [r^2h\,(2r - h)]}{R^2H - r^2h}\right\}$$

30. Shell of spherical segment

$$V = \pi\left[H^2\left(R - \frac{H}{3}\right) - h^2\left(r - \frac{h}{3}\right)\right]$$

$$B = \frac{3}{4}\left[\frac{\left(R - \frac{H}{3}\right)\frac{H^2\,(2R - H)^2}{3R - H} - \left(r - \frac{h}{3}\right)\frac{h^2\,(2r - h)^2}{3r - h}}{H^2\left(R - \frac{H}{3}\right) - h^2\left(r - \frac{h}{3}\right)}\right]$$

31. Circular hole through sphere

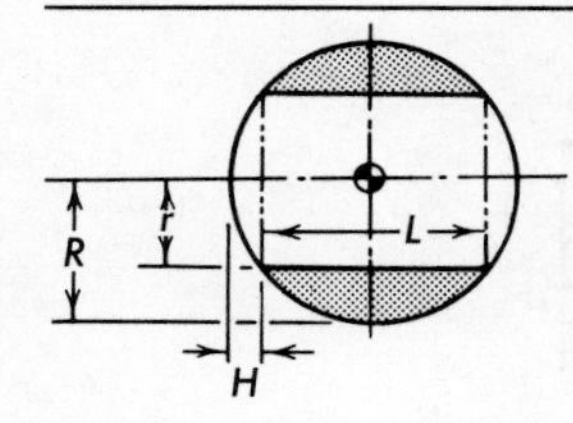

$$V = \pi\left[r^2L + 2H^2\left(R - \frac{H}{3}\right)\right] \qquad H = R - \sqrt{R^2 - r^2}$$

$$L = 2(R - H)$$

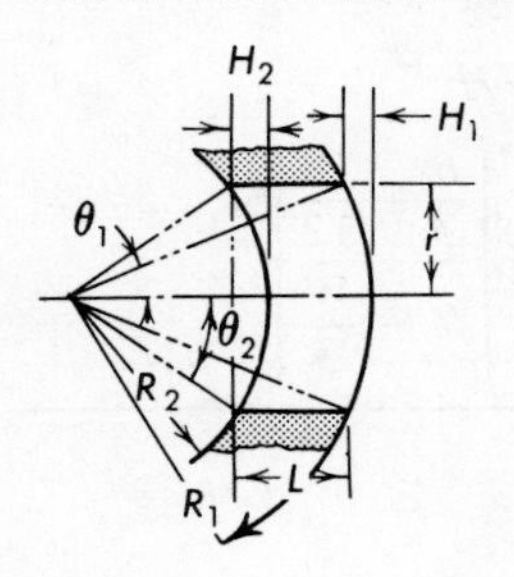

32. Circular hole through hollow sphere

$$V = \pi\left\{r^2L + H_1\left(R_1 - \frac{H_1}{3}\right) - H_2{}^2\left(R_2 - \frac{H_2}{3}\right)\right\}$$

$\sin\theta_1 = r/R_1 \qquad \sin\theta_2 = r/R_2 \qquad H = R\,(1 - \cos\theta)$

33. Spherical zone

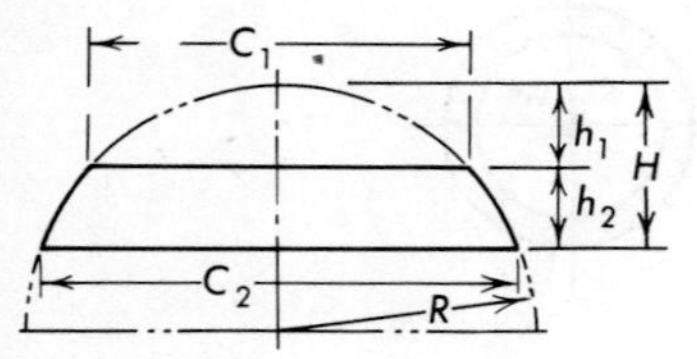

$$V = \pi\left\{\left[H^2\left(R - \frac{H}{3}\right)\right] - \left[h_1^2\left(R - \frac{h_1}{3}\right)\right]\right\}$$

$$V = \frac{\pi h_2}{6}\left[\frac{3}{4}C_1^2 + \frac{3}{4}C_2^2 + h_2^2\right]$$

34. Conical hole through spherical shell

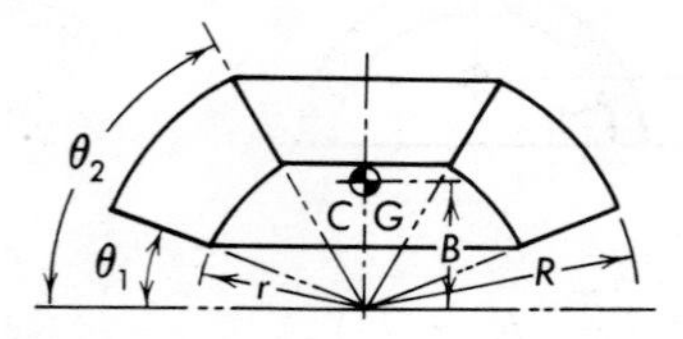

$$V = \frac{2\pi}{3}(R^3 - r^3)(\sin\theta_2 - \sin\theta_1)$$

$$B = \frac{0.375\,(R^4 - r^4)(\sin\theta_2 + \sin\theta_1)}{R^3 - r^3}$$

35. Torus

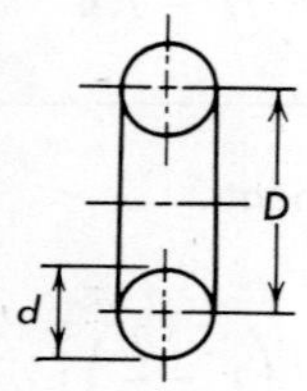

$$V = \frac{1}{4}\pi^2 d^2 D = 2.467 d^2 D$$

36. Hollow torus

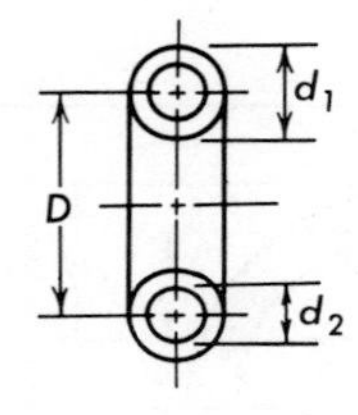

$$V = \frac{1}{4}\pi^2 D\,(d_1^2 - d_2^2)$$

37. Bevel ring

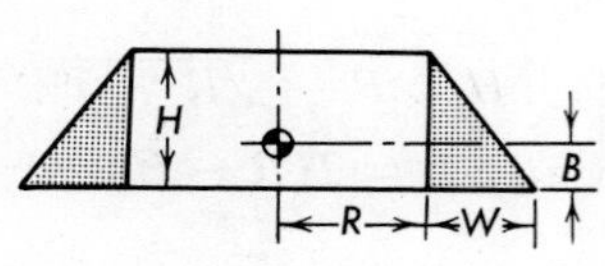

$$V = \pi\left(R + \frac{1}{3}W\right)WH$$

$$B = H\left[\frac{\frac{R}{3} + \frac{W}{12}}{R + \frac{W}{3}}\right]$$

38. Bevel ring

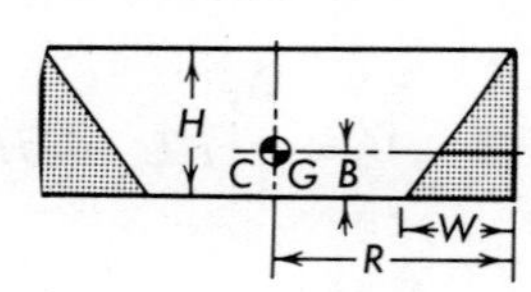

$$B > \frac{H}{3}$$

$$V = \pi\left(R - \frac{1}{3}W\right)WH$$

$$B = H\left[\frac{\frac{R}{3} - \frac{W}{12}}{R - \frac{W}{3}}\right]$$

39. Quarter torus

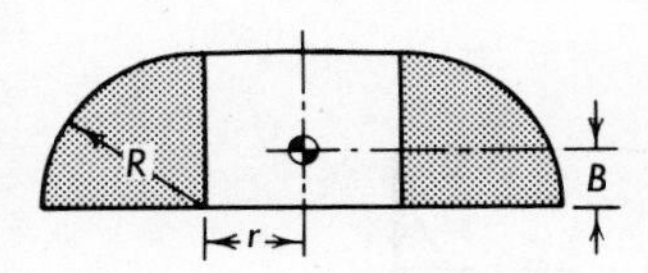

$$B < 0.4244R$$

$$V = \frac{\pi^2 R^2}{2}\left(r + \frac{4R}{3\pi}\right) = 4.9348R^2\,(r + 0.4244R)$$

$$B = \frac{4R}{3\pi}\left[\frac{r + \dfrac{3R}{8}}{r + \dfrac{4R}{3\pi}}\right] = \frac{0.4244Rr + 0.1592R^2}{r + 0.4244R}$$

40. Quarter torus

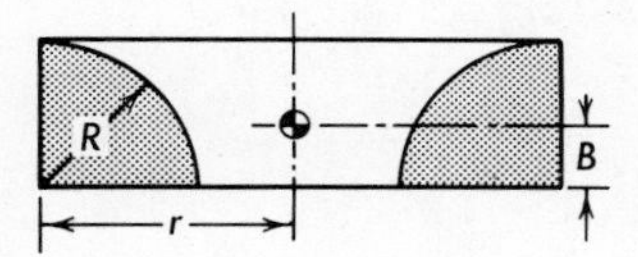

$$V = \frac{\pi^2 R^2}{2}\left[r - \frac{4R}{3\pi}\right]$$

$$B = \frac{4R}{3\pi}\left[\frac{r - \dfrac{3R}{8}}{r - \dfrac{4R}{3\pi}}\right]$$

41. Curved shell ring

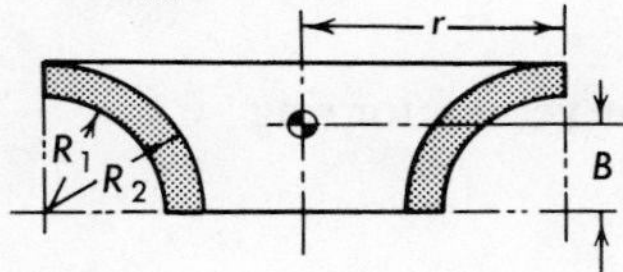

$$V = 2\pi\left\{r - \frac{4}{3\pi}\left[\frac{R_2^3 - R_1^3}{R_2^2 - R_1^2}\right]\right\}\frac{\pi}{4}\,(R_2^2 - R_1^2)$$

$$B = \frac{4}{3\pi}\left[\frac{R_2^3\left(r - \dfrac{3}{8}R_2\right) - R_1^3\left(r - \dfrac{3}{8}R_1\right)}{(R_2^2 - R_1^2)\left\{r - \dfrac{4}{3\pi}\left[\dfrac{R_2^3 - R_1^3}{R_2^2 - R_1^2}\right]\right\}}\right]$$

42. Curved shell ring

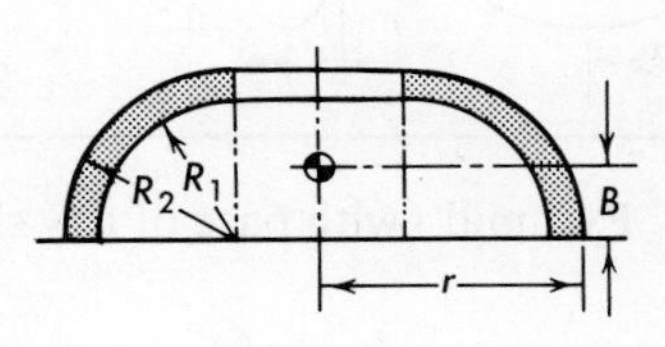

$$V = \frac{\pi^2}{2}\left[r(R_2^2 - R_1^2) + \frac{4}{3\pi}\,(R_2^3 - R_1^3)\right]$$

$$B = \frac{2}{\pi}\left[\frac{\dfrac{2r}{3}\,(R_2^3 - R_1^3) + \dfrac{1}{4}\,(R_2^4 - R_1^4)}{r(R_2^2 - R_1^2) + \dfrac{4}{3\pi}\,(R_2^3 - R_1^3)}\right]$$

43. Fillet ring

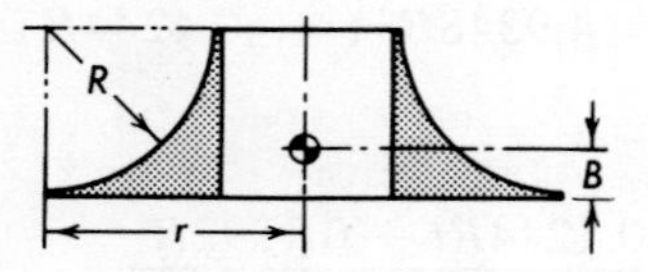

$$V = 2\pi R^2\left[\left(1-\frac{\pi}{4}\right)r - \frac{R}{6}\right]$$

$$B = R\left[\frac{\left(\frac{5}{6}-\frac{\pi}{4}\right)r - \frac{R}{24}}{\left(1-\frac{\pi}{4}\right)r - \frac{R}{6}}\right]$$

44. Fillet ring

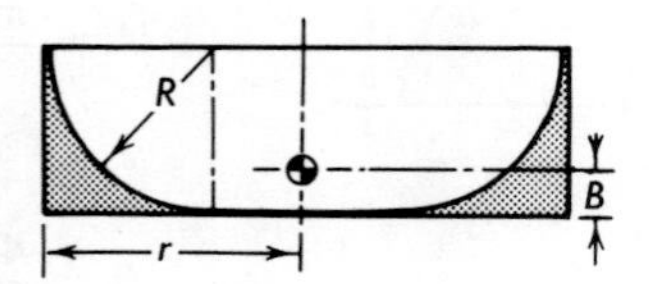

$$V = 2\pi R^2\left[\left(1-\frac{\pi}{4}\right)r - \left(\frac{5}{6}-\frac{\pi}{4}\right)R\right]$$

$$B = R\left[\frac{\left(\frac{5}{6}-\frac{\pi}{4}\right)r - \left(\frac{19}{24}-\frac{\pi}{4}\right)R}{\left(1-\frac{\pi}{4}\right)r - \left(\frac{5}{6}-\frac{\pi}{4}\right)R}\right]$$

45. Curved-sector ring

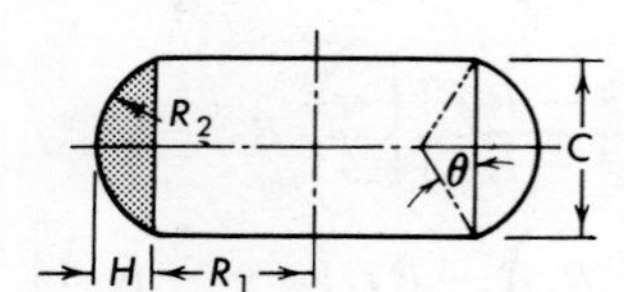

$$V = 2\pi R_2^2\left[R_1 + \left(\frac{4\sin 3\theta}{6\theta - 3\sin 2\theta} - \cos\theta\right)R_2\right]\bigl[\theta - 0.5\sin 2\theta$$

46. Ellipsoidal cylinder

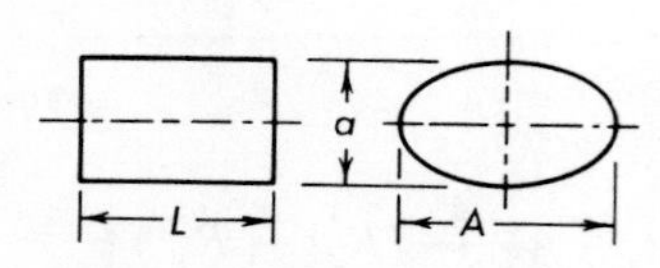

$$V = \frac{\pi}{4}AaL$$

47. Ellipsoid

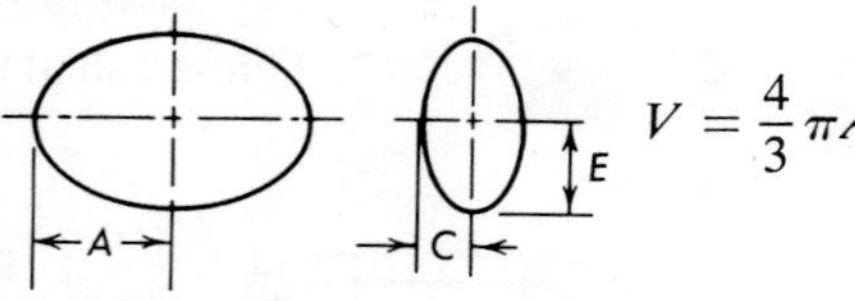

$$V = \frac{4}{3}\pi ACE$$

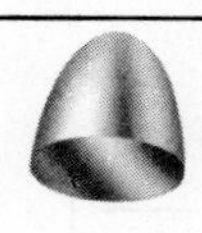

48. Paraboloid

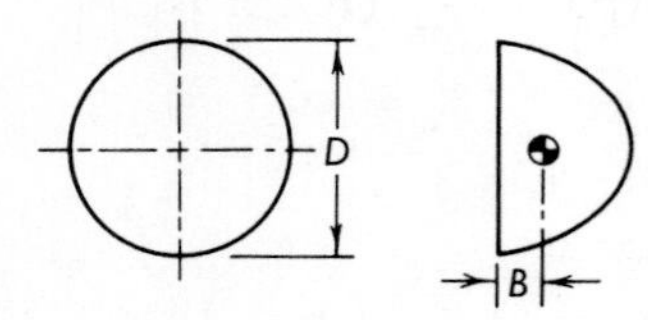

$$V = \frac{\pi}{8}HD^2 \qquad B = \frac{1}{3}H$$

49. Pyramid (with base of any shape)

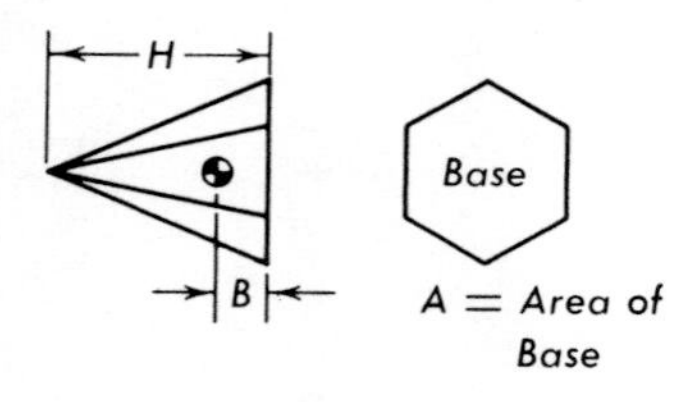

A = Area of Base $\quad V = \frac{1}{3}AH \quad B = \frac{1}{4}H$

50. Frustum of pyramid (with base of any shape)

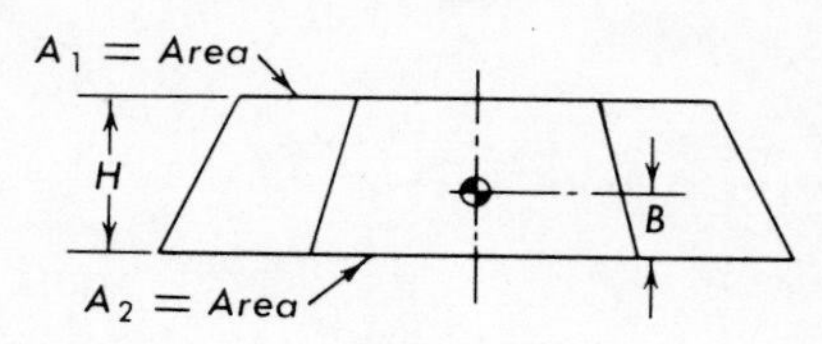

$$V = \frac{1}{3} H \left(A_1 + \sqrt{A_1 A_2} + A_2\right)$$

$$B = \frac{H \left(A_1 + 2\sqrt{A_1 A_2} + 3A_2\right)}{4 \left(A_1 + \sqrt{A_1 A_2} + A_2\right)}$$

51. Cone

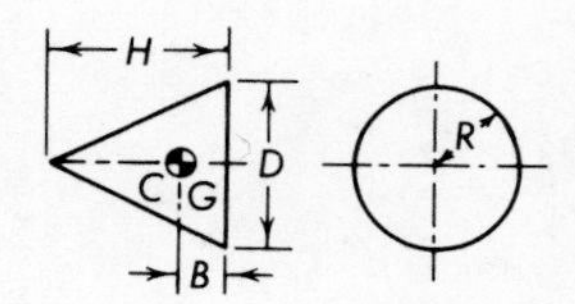

$$V = \frac{\pi}{12} D^2 H \qquad B = \frac{1}{4} H$$

Area of conical surface (right cone) = ½ (circumference of base) × (slant height)

52. Frustum of cone

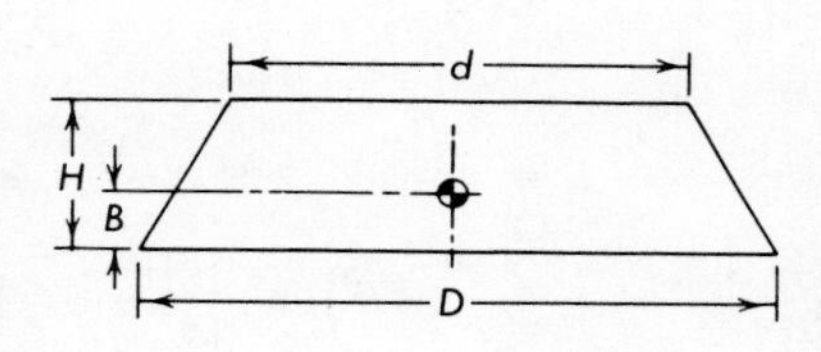

$$V = \frac{\pi}{12} H \left(D^2 + Dd + d^2\right)$$

$$B = \frac{H \left(D^2 + 2Dd + 3d^2\right)}{4 \left(D^2 + Dd + d^2\right)}$$

53. Frustum of hollow cone

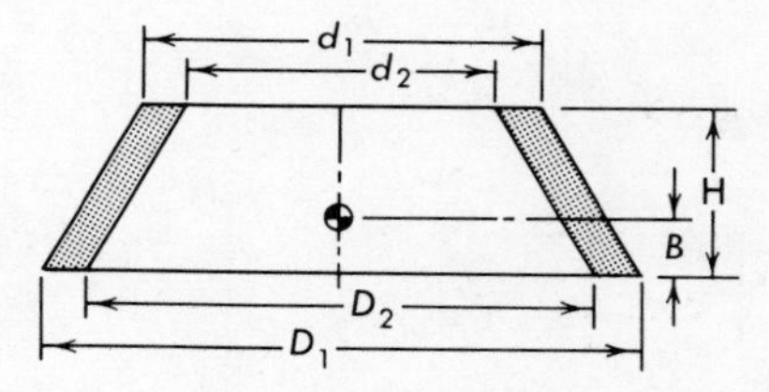

$$V = 0.2618H \left[\left(D_1^2 + D_1 d_1 + d_1^2\right) - \left(D_2^2 + D_2 d_2 + d_2^2\right)\right]$$

Appendix IV conversion factors and tables

The number in parentheses following a value in the table indicates the power of 10 by which this value should be multiplied. Thus, 6.214(-6) means 6.214×10^{-6}.

1. Length Equivalents

	cm	*in*	*ft*	*m*	*mi**
cm	1	3.937(-1)	3.281(-2)	1.0(-2)	6.214(-6)
in	2.540	1	8.333(-2)	2.54(-2)	1.578(-5)
ft	3.048(+1)	1.2(+1)	1	3.048(-1)	1.894(-4)
m	1.0(+2)	3.937(+1)	3.281	1	6.214(-4)
mi	1.609(+5)	6.336(+4)	5.280(+3)	1.609(+3)	1

*mile

Additional Measures

Metric:
- 1 km = 10^3 m
- 1 mm = 10^{-3} m
- 1 μm = 10^{-6} m (micron)
- 1 Å = 10^{-10} m (angstrom)

English:
- 1 mil = 10^{-3} in.
- 1 yd = 3.0 ft
- 1 rod = 5.5 yd = 16.5 ft
- 1 furlong = 40 rod = 660 ft

2. Area Equivalents

	m^2	*in^2*	*ft^2*	*acres*	*mi^2*
m^2	1	1.55(+3)	1.076(+1)	2.471(-4)	3.861(-7)
in^2	6.452(-4)	1	6.944(-3)	1.594(-7)	2.491(-10)
ft^2	9.290(-2)	1.44(+2)	1	2.296(-5)	3.587(-8)
acres	4.047(+3)	6.273(+6)	4.356(+4)	1	1.562(-3)
mi^2	2.590(+6)	4.018(+9)	2.788(+7)	6.40(+2)	1

Additional Measures

1 hectare = 10^4 m^2

3. Volume Equivalents

	cm³	*in³*	*ft³*	*gal (U.S.)*
cm²	1	6.103(−2)	3.532(−5)	2.642(−4)
in²	1.639(+1)	1	5.787(−4)	4.329(−3)
ft³	2.832(+4)	1.728(+3)	1	7.481
gal	3.785(+3)	2.31(+2)	1.337(−1)	1

Additional Measures

Metric: 1 liter = 10^3 cm³
1 m³ = 10^6 cm³

English: 1 quart = 0.250 gal (U.S.)
1 bushel = 9.309 gal (U.S.)
1 barrel = 42 gal (U.S.) (petroleum measure only)
1 imperial gal = 1.20 gal (U.S.) approx.
1 board-foot = 144 in³
1 chord (wood) = 128 ft³

4. Mass Equivalents

	kg	*slug*	*lb_m^**	*g*
Kg	1	6.85(−2)	2.205	1.0(+3)
slug	1.46(+1)	1	3.22(+1)	1.46(+4)
lb_m	4.54(−1)	3.11(−2)	1	4.54(+2)
g	1.0(−3)	6.85(−5)	2.205(−3)	1

*not recommended

5. Force Equivalents

	*N***	*lb*†	*dyne*	*kg_f**	*g_f**	*poundal**
N	1	2.248(−1)	1.0(+5)	1.019(−1)	1.019(+2)	7.234
lb_f	4.448	1	4.448(+5)	4.54(−1)	4.54(+2)	3.217(+1)
dyn	1.0(−5)	2.248(−6)	1	1.02(−6)	1.02(−3)	7.233(−5)
Kg_f	9.807	2.205	9.807(+5)	1	1.0(+3)	7.093(+1)
g_f	9.807(−3)	2.205(−3)	9.807(+2)	1.0(−3)	1	7.093(−2)
poundal	1.382(−1)	3.108(−2)	1.383(+4)	1.410(−2)	1.410(+1)	1

*not recommended **Newton †Avoirdupois

Additional Measures

1 metric ton = 10^3 kg_f = 2.205 × 10^3 lb_f
1 pound troy = 0.8229 lb_f
1 oz† = 6.25 × 10^{-2} lb_f
1 oz troy = 6.857 × 10^{-2} lb_f

6. Velocity and Acceleration Equivalents

Velocity

	cm/s	*ft/s*	*mi/h (mph)*	*km/h*
cm/s	1	3.28(−2)	2.237(−2)	3.60(−2)
ft/s	3.048(+1)	1	6.818(−1)	1.097
mi/h	4.470(+1)	1.467	1	1.609
km lb	2.778(+1)	9.113(−1)	6.214(−1)	1

*Standard acceleration of gravity

Acceleration

	cm/s²	*ft/s²*	$\overline{g}$*
cm/s²	1	3.281(−2)	1.019(−3)
ft/s²	3.048(+1)	1	3.109(−2)
$\overline{g}$	9.807(+2)	3.217(+1)	1

Additional Measures

1 knot = 1.152 miles/hr

7. Pressure Equivalents

						Head†	
	cyn/cm²	*N/m²*	*lb$_f$/in² (psi)*	*lb/ft² (psf)*	*atm**	*in (Hg)*	*ft (H_2O)*
dyn/cm²	1	1.0(−1)	1.45(−5)	2.089(−3)	9.869(−7)	2.953(−5)	3.349(−5)
N/m²	1.0(+1)	1	1.45(−4)	2.089(−2)	9.869(−6)	2.953(−4)	3.349(−4)
lb/in²	6.895(+4)	6.895(+3)	1	1.44(+2)	6.805(−2)	2.036	2.309
ft/lb²	4.788(+2)	4.788(+1)	6.944(−3)	1	4.725(−4)	1.414(−2)	1.603(−2)
atm	1.013(+6)	1.013(+5)	1.47(+1)	2.116(+3)	1	2.992(+1)	3.393(+1)
in (Hg)	3.386(+4)	3.386(+3)	4.912(−1)	7.073(+1)	3.342(−2)	1	1.134
ft (H_2D)	2.986(+4)	2.986(+3)	4.331(−1)	6.237(+1)	2.947(−2)	8.819(−1)	1

*Standard atmospheric pressure.
†At std. gravity and 0°C for Hg. 15°C for H_2O.

Additional Measures

1 bar = 1 N/cm²
1 pascal = 1 N/m²

8. Work and Energy Equivalents

	J/s	*ft-lb$_f$*	*W-h*	*Btu***	*Kcal*†	*kg-m*
J	1	7.376(−1)	2.778(−4)	9.478(−4)	2.388(−4)	1.020(−1)
ft-lb$_f$	1.356	1	3.766(−4)	1.285(−3)	3.238(−4)	1.383(−1)
W-h	3.60(+3)	2.655(+3)	1	3.412	8.599(−1)	3.671(+2)
Btu	1.055(+3)	7.782(+2)	2.931(−1)	1	2.520(−1)	1.076(+2)
Kcal	4.187(+3)	3.088(+3)	1.163	3.968	1	4.269(+2)
Kg-m	9.807	7.233	2.724(−3)	9.295(−3)	2.342(−3)	1

*Joule **British thermal unit † = kilocalorie

Additional Measures

1 Newton-meter = 1 J
1 erg = 1 dyne-cm = 10^{-7} J
1 cal = 10^{-3} kcal
1 therm = 10^{-5} Btu

9. Power Equivalents

	J/s	*ft-lb$_f$/s*	*hp*††	kW	Btu/h
J/s	1	7.376(−1)	1.341(−3)	1.0(−3)	3.412
ft-lb$_f$/s	1.356	1	1.818(−3)	1.356(−3)	4.626
hp	7.457(+2)	5.50(+2)	1	7.457(−1)	2.545(+3)
kW	1.0(+3)	7.376(+2)	1.341	1	3.412(+3)
Btu/h	2.931(−1)	2.162(−1)	3.930(−4)	2.931(−4)	1

††Horsepower

Additional Measures

1 W = 10^{-3} kW
1 cal/s = 14.29 Btu/h
1 poncelet - 100 kg-m/sec = 0.9807 kW
1 ton of refrigeration = 1.2×10^4 Btu/h

10. Time

1 week	7 days	168 hours	10,080 minutes	604,800 seconds
1 mean solar day			1440 minutes	86,400 seconds
1 calendar year	365 days	8760 hours	$5,256(10)^5$ minutes	$3.1536(10)^7$ seconds
1 tropical mean solar year	365.2422 days (basis of modern calendar)			

11. Temperature

$\Delta 1°$ Celsius (formerly Centigrade) (C) = $\Delta 1°$ Kelvin (K) = 1.8° Fahrenheit (F)
= 1.8° Rankine (R)

0°C = 273.15°K = 32°F = 491.67°R = 0°R
0°K = 273.15°C = 459.67°F

12. Electrical

1 coulomb	$1.036(10)^5$ faradays	0.1 abcoulomb	$2.998(10)^9$ statcoulombs
1 ampere		0.1 abampere	$2.998(10)^9$ statcoulombs
1 volt	10^3 millivolts	10^8 abvolts	$3.335(10)^{-3}$ statvolt
1 ohm	10^6 megohms	10^9 abohms	$1.112(10)^{-12}$ statohm
1 farad	10^6 microfarads	10^{-9} abfarads	$8.987(10)^{11}$ staffarads
1 henrie	10^3 millihenries	10^9 abhenries	$1.112(10)^{-12}$ stathenries
1 tesla	10^{-4} gauss		

THE GREEK ALPHABET

Α	α	Alpha	Ν	ν	Nu
Β	β	Bēta	Ξ	ξ	Xī
Γ	γ	Gamma	Ο	ο	Omicron
Δ	δ	Delta	Π	π	Pī
Ε	ϵ	Epsilon	Ρ	ρ	Rhō
Ζ	ζ	Zēta	Σ	σ	Sigma
Η	η	Eta	Τ	τ	Tau
Θ	θ	Thēta	Υ	υ	Upsilon
Ι	ι	Iōta	Φ	φ	Phī
Κ	κ	Kappa	Χ	χ	Chī
Λ	λ	Lambda	Ψ	ψ	Psī
Μ	μ	Mu	Ω	ω	Omega

DIMENSIONAL PREFIXES

Symbol	Prefix	Multiple
T	tera units	10^{12}
G	giga units	10^{9}
M	mega units	10^{6}
k	kilo units	10^{3}
h	hecto units	10^{2}
da	deca units	10^{1}
	units	10^{0}
d	deci units	10^{-1}
c	centi units	10^{-2}
m	milli units	10^{-3}
μ	micro units	10^{-6}
n	nano units	10^{-9}
p	pico units	10^{-12}
f	femto units	10^{-15}
a	atto units	10^{-18}

COEFFICIENTS OF FRICTION

Average values

Surfaces	Static	Kinetic
Metals on wood	0.4 –0.63	0.35–0.60
Wood on wood	0.3 –0.5	0.25–0.4
Leather on wood	0.38–0.45	0.3 –0.35
Iron on iron (wrought)	0.4 –0.5	0.4 –0.5
Glass on glass	0.23–0.25	0.20–0.25
Leather on glass	0.35–0.38	0.33–0.35
Wood on glass	0.35–0.40	0.28–0.31
Wood on sheet iron	0.43–0.50	0.38–0.45
Leather on sheet iron	0.45–0.50	0.35–0.40
Brass on wrought iron	0.35–0.45	0.30–0.35
Babbitt on steel	0.35–0.40	0.30–0.35
Steel on ice	0.03–0.04	0.03–0.04

SPECIFIC GRAVITIES AND SPECIFIC WEIGHTS

Material	Average specific gravity	Average specific weight, lb_f/ft^3	Material	Average specific gravity	Average specific weight, lb_f/ft^3
Acid, sulphuric, 87%	1.80	112	*Iron, grey cast*	7.10	450
Air, S.T.P.	0.001293	0.0806	*Iron, wrought*	7.75	480
Alcohol, ethyl	0.790	49			
Aluminum, cast	2.65	165	*Kerosene*	0.80	50
Asbestos	2.5	153			
Ash, white	0.67	42	*Lead*	11.34	710
Ashes, cinders	0.68	44	*Leather*	0.94	59
Asphaltum	1.3	81	*Limestone, solid*	2.70	168
			Limestone, crushed	1.50	95
Babbitt metal, soft	10.25	625			
Basalt, granite	1.50	96	*Mahogany*	0.70	44
Brass, cast-rolled	8.50	534	*Manganese*	7.42	475
Brick, common	1.90	119	*Marble*	2.70	166
Bronze, 7.9 to 14% S_n	8.1	509	*Mercury*	13.56	845
			Monel metal, rolled	8.97	555
Cedar, white, red	0.35	22			
Cement, portland, bags	1.44	90	*Nickel*	8.90	558
Chalk	2.25	140			
Clay, dry	1.00	63	*Oak, white*	0.77	48
Clay, loose, wet	1.75	110	*Oil, lubricating*	0.91	57
Coal, anthracite, solid	1.60	95			
Coal, bituminous, solid	1.35	85	*Paper*	0.92	58
Concrete, gravel, sand	2.3	142	*Paraffin*	0.90	56
Copper, cast, rolled	8.90	556	*Petroleum, crude*	0.88	55
Cork	0.24	15	*Pine, white*	0.43	27
Cotton, flax, hemp	1.48	93	*Platinum*	21.5	1330
Copper ore	4.2	262			
			Redwood, California	0.42	26
Earth	1.75	105	*Rubber*	1.25	78
Fir, Douglas	0.50	32	*Sand, loose, wet*	1.90	120
Flour, loose	0.45	28	*Sandstone, solid*	2.30	144
			Sea water	1.03	64
Gasoline	0.70	44	*Silver*	10.5	655
Glass, crown	2.60	161	*Steel, structural*	7.90	490
Glass, flint	3.30	205	*Sulphur*	2.00	125
Glycerine	1.25	78	*Teak, African*	0.99	62
Gold, cast-hammered	19.3	1205	*Tin*	7.30	456
Granite, solid	2.70	172	*Tungsten*	19.22	1200
Graphite	1.67	135	*Turpentine*	0.865	54
Gravel, loose, wet	1.68	105			
			Water, 4°C	1.00	62.4
Hickory	0.77	48	*Water, snow, fresh fallen*	0.125	8.0
Ice	0.91	57			
			Zinc	7.14	445

Note: The value for the specific weight of water which is usually used in problem solutions is 62.4 lb_f/ft^3 or 8.34 lb_f/gal.

LOGARITHMS

Natural Numbers	0	1	2	3	4	5	6	7	8	9	Proportional Parts								
											1	2	3	4	5	6	7	8	9
10	0000	0043	0086	0128	0170	0212	0253	0294	0334	0374	4	8	12	17	21	25	29	33	37
11	0414	0453	0492	0531	0569	0607	0645	0682	0719	0755	4	8	11	15	19	23	26	30	34
12	0792	0828	0864	0899	0934	0969	1004	1038	1072	1106	3	7	10	14	17	21	24	28	31
13	1139	1173	1206	1239	1271	1303	1335	1367	1399	1430	3	6	10	13	16	19	23	26	29
14	1461	1492	1523	1553	1584	1614	1644	1673	1703	1732	3	6	9	12	15	18	21	24	27
15	1761	1790	1818	1847	1875	1903	1931	1959	1987	2014	3	6	8	11	14	17	20	22	25
16	2041	2068	2095	2122	2148	2175	2201	2227	2253	2279	3	5	8	11	13	16	18	21	24
17	2304	2330	2355	2380	2405	2430	2455	2480	2504	2529	2	5	7	10	12	15	17	20	22
18	2553	2577	2601	2625	2648	2672	2695	2718	2742	2765	2	5	7	9	12	14	16	19	21
19	2788	2810	2833	2856	2878	2900	2923	2945	2967	2989	2	4	7	9	11	13	16	18	20
20	3010	3032	3054	3075	3096	3118	3139	3160	3181	3201	2	4	6	8	11	13	15	17	19
21	3222	3243	3263	3284	3304	3324	3345	3365	3385	3404	2	4	6	8	10	12	14	16	18
22	3424	3444	3464	3483	3502	3522	3541	3560	3579	3598	2	4	6	8	10	12	14	15	17
23	3617	3636	3655	3674	3692	3711	3729	3747	3766	3784	2	4	6	7	9	11	13	15	17
24	3802	3820	3838	3856	3874	3892	3909	3927	3945	3962	2	4	5	7	9	11	12	14	16
25	3979	3997	4014	4031	4048	4065	4082	4099	4116	4133	2	3	5	7	9	10	12	14	15
26	4150	4166	4183	4200	4216	4232	4249	4265	4281	4298	2	3	5	7	8	10	11	13	15
27	4314	4330	4346	4362	4378	4393	4409	4425	4440	4456	2	3	5	6	8	9	11	13	14
28	4472	4487	4502	4518	4533	4548	4564	4579	4594	4609	2	3	5	6	8	9	11	12	14
29	4624	4639	4654	4669	4683	4698	4713	4728	4742	4757	1	3	4	6	7	9	10	12	13
30	4771	4786	4800	4814	4829	4843	4857	4871	4886	4900	1	3	4	6	7	9	10	11	13
31	4914	4928	4942	4955	4969	4983	4997	5011	5024	5038	1	3	4	6	7	8	10	11	12
32	5051	5065	5079	5092	5105	5119	5132	5145	5159	5172	1	3	4	5	7	8	9	11	12
33	5185	5198	5211	5224	5237	5250	5263	5276	5289	5302	1	3	4	5	6	8	9	10	12
34	5315	5328	5340	5353	5366	5378	5391	5403	5416	5428	1	3	4	5	6	8	9	10	11
35	5441	5453	5465	5478	5490	5502	5514	5527	5539	5551	1	2	4	5	6	7	9	10	11
36	5563	5575	5587	5599	5611	5623	5635	5647	5658	5670	1	2	4	5	6	7	8	10	11
37	5682	5694	5705	5717	5729	5740	5752	5763	5775	5786	1	2	3	5	6	7	8	9	10
38	5798	5809	5821	5832	5843	5855	5866	5877	5888	5899	1	2	3	5	6	7	8	9	10
39	5911	5922	5933	5944	5955	5966	5977	5988	5999	6010	1	2	3	4	5	7	8	9	10
40	6021	6031	6042	6053	6064	6075	6085	6096	6107	6117	1	2	3	4	5	6	8	9	10
41	6128	6138	6149	6160	6170	6180	6191	6201	6212	6222	1	2	3	4	5	6	7	8	9
42	6232	6243	6253	6263	6274	6284	6294	6304	6314	6325	1	2	3	4	5	6	7	8	9
43	6335	6345	6355	6365	6375	6385	6395	6405	6415	6425	1	2	3	4	5	6	7	8	9
44	6435	6444	6454	6464	6474	6484	6493	6503	6513	6522	1	2	3	4	5	6	7	8	9
45	6532	6542	6551	6561	6571	6580	6590	6599	6609	6618	1	2	3	4	5	6	7	8	9
46	6628	6637	6646	6656	6665	6675	6684	6693	6702	6712	1	2	3	4	5	6	7	7	8
47	6721	6730	6739	6749	6758	6767	6776	6785	6794	6803	1	2	3	4	5	5	6	7	8
48	6812	6821	6830	6839	6848	6857	6866	6875	6884	6893	1	2	3	4	4	5	6	7	8
49	6902	6911	6920	6928	6937	6946	6955	6964	6972	6981	1	2	3	4	4	5	6	7	8
50	6990	6998	7007	7016	7024	7033	7042	7050	7059	7067	1	2	3	3	4	5	6	7	8
51	7076	7084	7093	7101	7110	7118	7126	7135	7143	7152	1	2	3	3	4	5	6	7	8
52	7160	7168	7177	7185	7193	7202	7210	7218	7226	7235	1	2	2	3	4	5	6	7	7
53	7243	7251	7259	7267	7275	7284	7292	7300	7308	7316	1	2	2	3	4	5	6	6	7
54	7324	7332	7340	7348	7356	7364	7372	7380	7388	7396	1	2	2	3	4	5	6	6	7

LOGARITHMS (continued)

Natural Numbers	0	1	2	3	4	5	6	7	8	9	Proportional Parts								
											1	2	3	4	5	6	7	8	9
55	7404	7412	7419	7427	7435	7443	7451	7459	7466	7474	1	2	2	3	4	5	5	6	7
56	7482	7490	7497	7505	7513	7520	7528	7536	7543	7551	1	2	2	3	4	5	5	6	7
57	7559	7566	7574	7582	7589	7597	7604	7612	7619	7627	1	2	2	3	4	5	5	6	7
58	7634	7642	7649	7657	7664	7672	7679	7686	7694	7701	1	1	2	3	4	4	5	6	7
59	7709	7716	7723	7731	7738	7745	7752	7760	7767	7774	1	1	2	3	4	4	5	6	7
60	7782	7789	7796	7803	7810	7818	7825	7832	7839	7846	1	1	2	3	4	4	5	6	6
61	7853	7860	7868	7875	7882	7889	7896	7903	7910	7917	1	1	2	3	4	4	5	6	6
62	7924	7931	7938	7945	7952	7959	7966	7973	7980	7987	1	1	2	3	3	4	5	6	6
63	7993	8000	8007	8014	8021	8028	8035	8041	8048	8055	1	1	2	3	3	4	5	5	6
64	8062	8069	8075	8082	8089	8096	8102	8109	8116	8122	1	1	2	3	3	4	5	5	6
65	8129	8136	8142	8149	8156	8162	8169	8176	8182	8189	1	1	2	3	3	4	5	5	6
66	8195	8202	8209	8215	8222	8228	8235	8241	8248	8254	1	1	2	3	3	4	5	5	6
67	8261	8267	8274	8280	8287	8293	8299	8306	8312	8319	1	1	2	3	3	4	5	5	6
68	8325	8331	8338	8344	8351	8357	8363	8370	8376	8382	1	1	2	3	3	4	4	5	6
69	8388	8395	8401	8407	8414	8420	8426	8432	8439	8445	1	1	2	2	3	4	4	5	6
70	8451	8457	8463	8470	8476	8482	8488	8494	8500	8506	1	1	2	2	3	4	4	5	6
71	8513	8519	8525	8531	8537	8543	8549	8555	8561	8567	1	1	2	2	3	4	4	5	5
72	8573	8579	8585	8591	8597	8603	8609	8615	8621	8627	1	1	2	2	3	4	4	5	5
73	8633	8639	8645	8651	8657	8663	8669	8675	8681	8686	1	1	2	2	3	4	4	5	5
74	8692	8698	8704	8710	8716	8722	8727	8733	8739	8745	1	1	2	2	3	4	4	5	5
75	8751	8756	8762	8768	8774	8779	8785	8791	8797	8802	1	1	2	2	3	3	4	5	5
76	8808	8814	8820	8825	8831	8837	8842	8848	8854	8859	1	1	2	2	3	3	4	5	5
77	8865	8871	8876	8882	8887	8893	8899	8904	8910	8915	1	1	2	2	3	3	4	4	5
78	8921	8927	8932	8938	8943	8949	8954	8960	8965	8971	1	1	2	2	3	3	4	4	5
79	8976	8982	8987	8993	8998	9004	9009	9015	9020	9026	1	1	2	2	3	3	4	4	5
80	9031	9036	9042	9047	9053	9058	9063	9069	9074	9079	1	1	2	2	3	3	4	4	5
81	9085	9090	9096	9101	9106	9112	9117	9122	9128	9133	1	1	2	2	3	3	4	4	5
82	9138	9143	9149	9154	9159	9165	9170	9175	9180	9186	1	1	2	2	3	3	4	4	5
83	9191	9196	9201	9206	9212	9217	9222	9227	9232	9238	1	1	2	2	3	3	4	4	5
84	9243	9248	9253	9258	9263	9269	9274	9279	9284	9289	1	1	2	2	3	3	4	4	5
85	9294	9299	9304	9309	9315	9320	9325	9330	9335	9340	1	1	2	2	3	3	4	4	5
86	9345	9350	9355	9360	9365	9370	9375	9380	9385	9390	1	1	2	2	3	3	4	4	5
87	9395	9400	9405	9410	9415	9420	9425	9430	9435	9440	0	1	1	2	2	3	3	4	4
88	9445	9450	9455	9460	9465	9469	9474	9479	9484	9489	0	1	1	2	2	3	3	4	4
89	9494	9499	9504	9509	9513	9518	9523	9528	9533	9538	0	1	1	2	2	3	3	4	4
90	9542	9547	9552	9557	9562	9566	9571	9576	9581	9586	0	1	1	2	2	3	3	4	4
91	9590	9595	9600	9605	9609	9614	9619	9624	9628	9633	0	1	1	2	2	3	3	4	4
92	9638	9643	9647	9652	9657	9661	9666	9671	9675	9680	0	1	1	2	2	3	3	4	4
93	9685	9689	9694	9699	9703	9708	9713	9717	9722	9727	0	1	1	2	2	3	3	4	4
94	9731	9736	9741	9745	9750	9754	9759	9763	9768	9773	0	1	1	2	2	3	3	4	4
95	9777	9782	9786	9791	9795	9800	9805	9809	9814	9818	0	1	1	2	2	3	3	4	4
96	9823	9827	9832	9836	9841	9845	9850	9854	9859	9863	0	1	1	2	2	3	3	4	4
97	9868	9872	9877	9881	9886	9890	9894	9899	9903	9908	0	1	1	2	2	3	3	4	4
98	9912	9917	9921	9926	9930	9934	9939	9943	9948	9952	0	1	1	2	2	3	3	4	4
99	9956	9961	9965	9969	9974	9978	9983	9987	9991	9996	0	1	1	2	2	3	3	3	4

TRIGONOMETRIC FUNCTIONS

Angle	*sin*	*cos*	*tan*	*cot*	*sec*	*csc*	
0°	0.0000	1.0000	0.0000	∞	1.0000	∞	90°
1°	0.0175	0.9998	0.0175	57.2900	1.0002	57.2987	89°
2°	0.0349	0.9994	0.0349	28.6363	1.0006	28.6537	88°
3°	0.0523	0.9986	0.0524	19.0811	1.0014	19.1073	87°
4°	0.0698	0.9976	0.0699	14.3007	1.0024	14.3356	86°
5°	0.0872	0.9962	0.0875	11.4301	1.0038	11.4737	85°
6°	0.1045	0.9945	0.1051	9.5144	1.0055	9.5668	84°
7°	0.1219	0.9925	0.1228	8.1443	1.0075	8.2055	83°
8°	0.1392	0.9903	0.1405	7.1154	1.0098	7.1853	82°
9°	0.1564	0.9877	0.1584	6.3138	1.0125	6.3925	81°
10°	0.1736	0.9848	0.1763	5.6713	1.0154	5.7588	80°
11°	0.1908	0.9816	0.1944	5.1446	1.0187	5.2408	79°
12°	0.2079	0.9781	0.2126	4.7046	1.0223	4.8097	78°
13°	0.2250	0.9744	0.2309	4.3315	1.0263	4.4454	77°
14°	0.2419	0.9703	0.2493	4.0108	1.0306	4.1336	76°
15°	0.2588	0.9659	0.2679	3.7321	1.0353	3.8637	75°
16°	0.2756	0.9613	0.2867	3.4874	1.0403	3.6280	74°
17°	0.2924	0.9563	0.3057	3.2709	1.0457	3.4203	73°
18°	0.3090	0.9511	0.3249	3.0777	1.0515	3.2361	72°
19°	0.3256	0.9455	0.3443	2.9042	1.0576	3.0716	71°
20°	0.3420	0.9397	0.3640	2.7475	1.0642	2.9238	70°
21°	0.3584	0.9336	0.3839	2.6051	1.0711	2.7904	69°
22°	0.3746	0.9272	0.4040	2.4751	1.0785	2.6695	68°
23°	0.3907	0.9205	0.4245	2.3559	1.0864	2.5593	67°
24°	0.4067	0.9135	0.4452	2.2460	1.0946	2.4586	66°
25°	0.4226	0.9063	0.4663	2.1445	1.1034	2.3662	65°
26°	0.4384	0.8988	0.4877	2.0503	1.1126	2.2812	64°
27°	0.4540	0.8910	0.5095	1.9626	1.1223	2.2027	63°
28°	0.4695	0.8829	0.5317	1.8807	1.1326	2.1301	62°
29°	0.4848	0.8746	0.5543	1.8040	1.1434	2.0627	61°
30°	0.5000	0.8660	0.5774	1.7321	1.1547	2.0000	60°
31°	0.5150	0.8572	0.6009	1.6643	1.1666	1.9416	59°
32°	0.5299	0.8480	0.6249	1.6003	1.1792	1.8871	58°
33°	0.5446	0.8387	0.6494	1.5399	1.1924	1.8361	57°
34°	0.5592	0.8290	0.6745	1.4826	1.2062	1.7883	56°
35°	0.5736	0.8192	0.7002	1.4281	1.2208	1.7434	55°
36°	0.5878	0.8090	0.7265	1.3764	1.2361	1.7013	54°
37°	0.6018	0.7986	0.7536	1.3270	1.2521	1.6616	53°
38°	0.6157	0.7880	0.7813	1.2799	1.2690	1.6243	52°
39°	0.6293	0.7771	0.8098	1.2349	1.2868	1.5890	51°
40°	0.6428	0.7660	0.8391	1.1918	1.3054	1.5557	50°
41°	0.6561	0.7547	0.8693	1.1504	1.3250	1.5243	49°
42°	0.6691	0.7431	0.9004	1.1106	1.3456	1.4945	48°
43°	0.6820	0.7314	0.9325	1.0724	1.3673	1.4663	47°
44°	0.6947	0.7193	0.9657	1.0355	1.3902	1.4396	46°
45°	0.7071	0.7071	1.0000	1.0000	1.4142	1.4142	45°
	cos	*sin*	*cot*	*tan*	*csc*	*sec*	*Angle*

Appendix V
answers to selected problems

1-1. e. $\chi = 5$
j. $\chi = -3$
n. $\chi = 4$

2-1. e. 2.7810
j. 7.7715
o. 6.8222
t. 3.6435
y. 7.8567

2-45. $8.51(10)^6$
2-20. $8.75(10)^1$
2-55. $3.79(10)^3$
2-60. $8.37(10)^5$
2-65. $4.35(10)^5$
2-70. $1.202(10)^6$
2-75. 1.095
2-80. $8.32(10)^3$
2-85. $6.06(10)^6$
2-90. $1.619(10)^6$
2-135. $2.53(10)^1$
2-140. $4.59(10)^1$
2-145. $4.25(10)^4$
2-150. $5.91(10)^2$
2-155. $2.77(10)^{-1}$
2-160. $3.88(10)^{-3}$
2-165. $1.275(10)^8$
2-170. 1.278
2-175. $2.21(10)^3$
2-180. $5.96(10)^3$
2-225. $1.350(10)^1$
2-230. $1.524(10)^{-2}$
2-235. $9.54(10)^{-1}$
2-240. $1.099(10)^{-1}$
2-245. $2.87(10)^{-6}$
2-250. $1.437(10)^{-3}$
2-255. $2.93(10)^{-7}$
2-260. $2.96(10)^4$
2-265. $5.07(10)^{-1}$
2-270. 5.64
2-355. $1.430(10)^2$
2-360. $4.46(10)^{-3}$
2-365. $2.02(10)^9$
2-370. $9.98(10)^{-1}$
2-375. 1.772
2-380. $5.27(10)^{-1}$
2-385. $3.62(10)^3$
2-390. $2.53(10)^1$
2-445. $1.079(10)^5$
2-450. $1.357(10)^6$
2-455. $8.12(10)^{-7}$
2-246. 2.08
2-465. $3.26(10)^1$
2-470. $3.68(10)^2$
2-475. $5.36(10)^{-2}$
2-480. $1.139(10)^3$
2-560. 0.978
2-565. 0.407
2-570. 0.669
2-575. 1.397
2-580. 1.028
2-585. 0.719
2-590. 1.034
2-595. 1.856
2-600. 1.061
2-605. 88.35°
2-610. 7.25°
2-650. 0.999
2-620. 31.8°
2-625. 29.56°
2-630. 0.602
2-635. 0.235
2-640. 0.897
2-645. 1.513
2-650. 0.1568
2-655. 1.168
2-660. b = 15.97
B = 23.4°
2-665. c = 4.09
B = 15.0°
2-670. a = 598
b = 1808
2-675. a = 677
c = 678
2-730. $1.113(10)^2$
2-735. 1.331
2-740. $1.048(10)^{-1}$
2-745. 1.0352
2-750. $6.97(10)^{11}$
2-755. $4.918(10)^{-1}$
2-760. 1.434
2-765. 1.0006
2-770. $3.85(10)^{-1}$
2-775. 1.444
2-780. 0.0439
2-785. 35.9
2-790. 3.89
2-795. $9.26(10)^{-1}$
2-800. 1.018
2-805. -0.250
2-810. 3.51

2-815. -6.70
2-820. 4.497
2-825. -0.0026
2-830. $3.11(10)^1$
2-835. $2.73(10)^{-2}$
2-840. $3.97(10)^4$
2-845. $-1.230(10)^3$
2-850. $1.706(10)^5$
2-855. $8.53(10)^3$
2-860. $8.98(10)^7$
2-865. $(9.42)(10)^{-2}$
2-870. $5.63(10)^1$
2-875. $5.43(10)^3$
2-880. $2.75(10)^{-3}$
2-885. 1.776
2-890. $3.26(10)^3$
2-895. $2.45(10)^{12}$
2-900. $7.39(10)^4$
2-905. 1.050
2-910. $6.84(10)^{-1}$
2-915. $3.35(10)^1$
2-920. $3.24(10)^{-1}$
2-925. $1.072(10)^{-9}$
2-930. a. 1.039
b. 1.579
c. 3.69
d. 5.395
e. 17.61
f. 28.42
g. 74.21
2-935. a. 4.81 + j 3.90
b. 2.41 + j 2.68
c. 8.58 + j 3.36
d. 0.88 + j 2.56
2-940. a. $218\underline{/317.4^\circ}$
b. $100\underline{/332.6^\circ}$
c. $0.00803\underline{/320.5^\circ}$
d. $3.65\underline{/327.4^\circ}$
4-5. $k = \dfrac{M^8L^3}{F^4\theta T}$
4-10. $k = \dfrac{FL^2}{Q^{\frac{1}{2}}M^3}$
4-15. a. $1.089(10)^6$ dynes/cm^2
b. $3.22(10)^1$ in Hg
4-20. $3.44(10)^8$ abhenries
4-25. a. $6.96(10)^4$ ft^3/hr
b. $5.47(10)^{-1}$ m^3/s
4-30. a. $2.07(10)^{-4}$ acre
b. $8.36(10)^{-1}$ m^2
4-35. $6.40(10)^{-2}$ ft^2
4-40. 1.274 in.
4-45. M = FL
4-50. a. KE = $2.26(10)^9$ ft lb_f
b. V = $7.5(10)^4$ ft/sec
c. M = 93 slugs
d. V = 145.6 mi/hr
4-55. a. m = $3.83(10)^2$ lbm
b. m = $1.189(10)^1$ slug
4-60. $m_1 = 5.97(10)^{24}$ kg
4-70. ⊙ = 95.4 □ ↑ ; Engineer correct
4-75. 0.21 m/s
4-80. 7.11 ohms
6-5. a. 25.05 in^3
b. 0.2076 lb
6-10. $1.3357(10)^4$ lb
6-15. a. $1.788(10)^3$ ft^2
b. $2.012(10)^4$ lb
6-20. 10.67 mm
6-25. a. 68°F
b. 98.6°F
c. 311°F
d. 1076°F
e. 15,852°F
f. $2.20(10)^5$ °F
g. 28.4°F
h. -40°F
i. -459°F
6-30. $480
6-35. R = 333 lb_f @ N 58.6 °E
6-40. P = 446 lb
6-45. $M_{188} = 0$
M_{102} = 395 lb-ft
M_{99} = 280 lb-ft
M_{177} = 780 lb-ft
6-52. R_R = 460 lb_f; R_L = 600 lb_f
6-57. P = 432 lb_f; R_L = 644 lb;
R_R = 440 lb
6-62. P. = 153.1 kg_f; N = 235 kg_f
6-67. AC = 1750 lb; R_B= 1405 lb_f
@ S 72.4°E

index